MW00736876

Field Strength at the center of a
circular coil $H = \dfrac{NI}{2r}$ amps/meter $(\times \cdot 01257 \text{ oersted})$

Flux density B at center of coil is therefore
$$B = \mu_0 H = \mu_0 NI/2r \text{ webers/meter}^2 \left(\times 10^4 \text{ gauss}\right)$$
p72 Inductance of RG58/u is $0.036 \,\mu H/ft$

ELECTRO-MAGNETIC INTERFERENCE REDUCTION IN ELECTRONIC SYSTEMS

Jeffrey P. Mills

PTR Prentice Hall, Englewood Cliffs, New Jersey 07632

Library of Congress Cataloging-in-Publication Data

Mills, Jeffrey P.
  Electromagnetic interference reduction in electronic systems /
Jeffrey P. Mills.
    p.   cm.
  ISBN 0-13-463902-2
  1. Electronic circuits—Noise.   2. Electromagnetic compatibility.
3. Shielding (Electricity)   I. Title.
TK7867.5.M55   1993
621.382'24—dc20

92-44408
CIP

To Sue

Editorial/Production Supervision: *The Wheetley Co., Inc.*
Buyer: *Mary Elizabeth McCartney*
Acquisitions Editor: *Karen Gettman*

The publisher offers discounts on this book when ordered
in bulk quantities. For more information, contact:

Corporate Sales Department
PTR Prentice Hall
113 Sylvan Avenue
Englewood Cliffs, New Jersey 07632

Phone: 201-592-2863
Fax: 201-592-2249

Printed in the United States of America
10 9   8   7   6   5   4   3   2   1

ISBN 0-13-463902-2

Prentice-Hall International (UK) Limited, *London*
Prentice-Hall of Australia Pty. Limited, *Sydney*
Prentice-Hall Canada Inc., *Toronto*
Prentice-Hall Hispanoamericana, S.A., *Mexico*
Prentice Hall of India Private Limited, *New Delhi*
Prentice-Hall of Japan, Inc., *Tokyo*
Simon & Schuster Asia Pts. Ltd., *Singapore*
Editors Prentice-Hall do Brasil, Ltda., *Rio De Janeiro*

CONTENTS

LIST OF ILLUSTRATIONS

PREFACE

With the proliferation of complex electronic equipment, electromagnetic compatibility—or EMC—has become an unavoidable concern of most electronic designers. In order to design a product with due consideration for EMC, the electrical engineer must understand certain specialized topics in electromagnetics. This book presents those topics in a complete and coherent manner, and it is suitable as a text for a one-semester senior or graduate-level course in electromagnetic compatibility. Prerequisite courses consist of one year of basic electromagnetic fields and transmission lines, normally included in the junior year of the undergraduate electrical engineering curriculum.

Most EMC concepts are familiar to engineers who specialize in electromagnetics. However, they are gathered from various elective courses offered throughout the undergraduate and graduate electrical engineering curriculum, and they are therefore unfamiliar to nonspecialists. Previously, EMC concepts have not been presented in an analytical fashion in a single textbook. Available books on this subject have typically addressed the experienced engineer who may not have used the material for some time and who must refresh his or her knowledge to fulfill the EMC objectives of a design project. Such books serve their intended purpose well but are unsuitable as college textbooks.

This book uses basic electromagnetic theory to deduce the design formulae required for effective positioning of components and shielding of cables and cabinets. It prepares a college student for specialization in EMC, whether for research or for consultation in industry. It also can give a student specializing in another area the knowledge required to design EMC into a product with less assistance from an EMC specialist. Treatment of the material is mathematical, correlatable with experimental results.

Electromagnetic theory begins with a discussion of unintentional coupling between nearby devices, for which electric and magnetic modes may be considered separately. Components on a chassis or printed-wiring card, and conductors in a cable, are studied. Circuit theory is correlated with field theory. Circuit theory should be familiar to all electrical engineers, but with this approach the components correspond to stray inductances and capacitances whose values are usually difficult to quantify. Field theory, on the other hand, requires knowledge of only the physical dimensions of the hardware, but is less familiar to those electrical engineers who do not specialize in electromagnetics. To achieve a sound conceptual understanding of the topic, circuit theory is first used to describe it qualitatively. Quantitative results are then obtained using field theory, primarily transmission lines.

Shielding of cables and its effects on electric and magnetic coupling are discussed. Various techniques are presented for grounding the cable shields and associated circuits. Stray current return paths, the nemesis of EMC control, are analyzed in connection with grounding a cable shield at multiple points. Control of these paths, using balanced circuits, common-mode impedance, and other means, is investigated. Field theory is again used to predict, quantitatively, the severity of the problem. Since safety grounding requirements often conflict with control of signal return paths, methods are presented that attempt to satisfy both objectives.

Radiated interference between devices is also treated analytically. The starting point, as with most topics in electromagnetics, is Maxwell's equations. Their solution, subject to the boundary conditions comprising a small dipole antenna, is presented and justified. Near and far-field terms are introduced. It is noted that any radiating system, intentional or unintentional, can be assumed to consist of a multitude of these dipole antennas, and therefore its field expression contains these near and far-field terms. Shielding effects on near and far electric and magnetic fields are discussed. Theory developed by S. A. Schelkunoff is used to decide the required thickness of a shield of inferior conductivity, such as a cabinet made of metal-impregnated plastic. Other methods of shielding, such as conductive coatings, are also presented.

Apertures in a shield, with and without leads passing through them, are investigated. Slot antenna theory is introduced to predict the effects of seams and other unavoidable discontinuities in cabinet walls. Permissible slot dimensions are determined for specified attenuation ratios and specified frequencies of emissions.

An aspect of EMC that has been severely overlooked is the consequences of insulated leads passing through a shield aperture. It is shown that without proper treatment a single DC power lead passing through a very small hole in a shield can render the shield ineffective. Methods for preserving the integrity of a shield, while allowing necessary low-frequency and high-frequency leads to pass through it, are discussed. Antenna theory is used here as required.

To apply the electromagnetic theory discussed in this book, the spectral content of any undesired emissions must be known. Spectral analysis of periodic and random signals is therefore discussed, to the extent necessary to characterize the emissions for EMC analysis. Broadband emissions are compared with narrowband emissions, and the nature of the interference normally caused by each type is discussed. A basic knowledge of statistics and random signals is assumed.

The necessity for EMC regulations is noted, and the present domestic and international regulations are discussed. Compliance with these regulations requires an understanding of the measurement techniques stated or implied in the rules. Various measurement techniques are compared, and their advantages and costs are weighed. Open-field sites are compared with shielded rooms and anechoic chambers, and the quality of construction required for a suitable measurement environment is examined.

No textbook on EMC would be complete without a justification for the need to consider EMC early in the design of electronic products. Comparative costs are identified with EMC strategies introduced at various stages of manufacture. Also cited are examples of EMC difficulties often incurred when faster, more sophisticated devices must be substituted for older, slower components that are no longer available.

With every development in electrical engineering, new potentials for electromagnetic interference arise. Knowing their possible consequences, future electrical engineers should be motivated to incorporate EMC into their designs. This book addresses typical designs of today and is intended to train the student to apply similar design principles to the products of the future.

I am deeply indebted to many associates and former students for their comments and advice. Particular thanks are expressed to John Barnes, Joseph LoCicero, Erwin Weber, and Kenneth Wyatt for their assistance in reviewing the manuscript. Finally, I am most grateful to my wife, Susan, for her assistance in typing the manuscript and for her encouragement throughout the preparation of this book.

JEFFREY P. MILLS

1

CONCERN FOR ELECTROMAGNETIC COMPATIBILITY

Electric and magnetic fields exist near any operating electric circuit. The current flowing in the circuit generates a magnetic field. For such a current to flow, there must be a potential difference, which produces an electric field. These fields can exist in any substance, including a vacuum. If they are strong enough, they make it possible for one electrical device to affect another. When this occurs unintentionally, it is known as *electromagnetic interference*, commonly abbreviated *EMI* and often simply called "noise."

Problems caused by EMI have been noticed for centuries. The first human experience with EMI probably involved the compass, which the Chinese used in the eleventh century. Although no one knew about electric currents at that time, it was true then as now that two nearby compasses would interfere with each other. It is also likely that natural phenomena such as lightning could have affected the magnetization of the lodestone used in such early compasses. With the limited scientific knowledge of that era, however, it is unlikely that anyone could explain the resulting erratic behavior of the compass.

Although our knowledge has expanded tremendously since then, so has the complexity of our inventions. Speeds of electronic circuits are increasing at an enormous rate. With this increased speed comes a greater likelihood of EMI. A design that was adequate five years ago might be unacceptable when used with the faster components available today. This can be a serious problem when older, slower components cease to be available. A faster component, designed to supersede an older one, may fail to operate correctly because of EMI.

It is very easy to overlook EMI in system design, though every electrical engineer knows its basics. When EMI is overlooked, a system may cause a malfunction in another apparatus located nearby, or the system itself may not operate correctly because its subsystems interfere with each other even though each functions properly by itself. Tersely, "everything checks but nothing works." In either case, the result is unsatisfactory, often requiring expensive redesign.

The results of an EMI problem may range from a simple annoyance to a major disaster. Noise from a personal computer may affect a nearby TV receiver. The result here might be a feud between two family members, or between two neighbors in an apartment building. At the other extreme, the control surfaces of some aircraft have been observed to move uncontrollably when subjected to strong electromagnetic fields. The consequences of a poor design such as this are obvious. Other examples of interference include the following:

- an electric motor causing "static" on a TV or radio receiver;
- a radio broadcast transmitter audible on a telephone line;
- an automotive ignition system causing noise on a nearby radio, possibly in the same vehicle.

In each case, the noise is usually coupled through electric and/or magnetic fields in space, which we call *radiated* coupling. However, another coupling path exists if two appliances share the same power source. One appliance may generate undesired high-frequency voltages on its power leads, which then appear on the power leads of the other. The second appliance may then malfunction due to this voltage. This is *conducted* coupling. So we must consider both radiated and conducted noise.

Webster[1] defines *compatibility* as "the capability of components (as of an electronic system) to function together." Electromagnetic compatibility, or *EMC*, refers to the capability of two or more electrical devices to operate simultaneously without mutual interference. A personal computer and its video monitor are examples of two devices in the same system that might interfere with each other. Even two parts of the same appliance that would operate correctly by themselves, such as two amplifier stages, might not operate satisfactorily together. Obviously, a basic EMC requirement is that no part of a circuit may interfere with another part. The engineer must consider this in system design but can easily test for it. If the complete system works correctly under all modes of operation, it can be assumed that all of its parts are compatible. Separate appliances that normally operate together also can be tested easily, although the number of combinations may grow large.

Interference between two unrelated systems is much more difficult to deal with from a design standpoint. Often a problem may be solved simply by moving the equipment cabinets farther apart, but sometimes this is not practical. So electromagnetic compatibility must be designed into each system. The designer has no knowledge of what kinds of other equipment will be located near his system. Thus, it is nearly impossible to predict actual interference to or from other equipment with

---

1. Phillip Babcock Gove, ed., *Webster's Third New International Dictionary* (Springfield, Mass.: Merriam-Webster, Inc., 1986), p. 463.

precision. Instead, we seek design guidelines that provide a reasonable degree of confidence for a typical installation. Two systems designed accordingly are then unlikely to interfere when separated by a reasonable distance.

For interference to occur, a source must *emit* noise (radiated or conducted) that affects a *susceptible* victim. At a given frequency, the source emissions may be deliberate or accidental. Also, the victim may be either intentionally or inadvertently susceptible to emissions at that frequency. If both are intentional, we do not have interference; we have a normal communication channel. Interference exists when the source or victim, or both, are unplanned. Examples of intentional and unintentional emitters and sensors appear in Figure 1-1.

Stray sources are usually much weaker than deliberate sources. Also, inadvertent receptors are usually much less sensitive than intended receptors. Thus, interference between a stray source and a stray receptor is common only when they are very close together. This may occur between two sub-assemblies in the same appliance. Otherwise, interference usually occurs only when either the source or the receiver is intended as such. In the first case, the victim is accidentally susceptible, whereas in the second case a stray source is emitting. Presumably, an intentional emitter or sensor cannot be altered. Therefore, to avoid interference, equipment should be designed not only to minimize stray emissions but also to tolerate inflicted noise. Fortunately, due to reciprocity, procedures that reduce emissions usually also reduce susceptibility at the same frequencies. One method is to weaken any fields near the equipment by adding a shield, as will be discussed later.

An apparatus that is susceptible to undesired noise will usually not perform satisfactorily in the presence of that noise. A knowledgeable customer will discover this and not be satisfied with the product. This is usually sufficient incentive for the manufacturer to design noise immunity into his product. On the other hand, an appliance may function acceptably itself while emitting stray noise that causes other

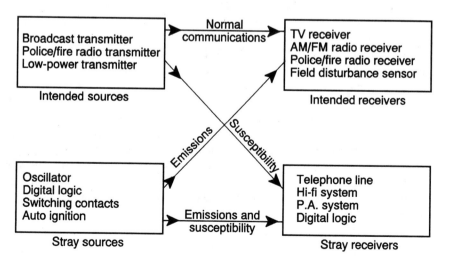

Figure 1-1   Intentional and unintentional EMI sources and sensors

equipment to malfunction. If someone else owns the affected equipment, then the user of the interfering source may not even be aware that his appliance is causing problems. Therefore, there would be less incentive to reduce stray emissions if the law did not require it.

For several decades, regulations have dictated that if any electronic device causes harmful interference, the user must shut it down. This has sometimes forced a user to cease using an appliance permanently. In one famous case, the user then sold his appliance, a coin-operated video game, to someone else. The noise did not disappear but only moved with the video game, and the problem persisted. Such noise is usually due to poor design, which is the manufacturer's responsibility, not the user's. To be fair to the user, every manufacturer should be banned from building unusable products. More recent rules have established absolute limits for both radiated and conducted emissions. A manufacturer must meet these limits before he can advertise or sell his product, so the responsibility is now his *and* the user's. These regulations will be discussed in Chapter 12.

Reduction of stray emissions and accidental susceptibility must be considered early in the design of a system; otherwise, expensive redesign may become necessary. Spacing between two conductors on a printed wiring board can be controlled with one keystroke at the time of layout. After manufacture of the board, any changes must be hand-wired on every such board by skilled technicians. Problems are even worse if the customer has already purchased and received the equipment, which the manufacturer must then recall. It may be difficult to locate all faulty boards, and the manufacturer's reputation may suffer.

The first step in EMC design is to investigate the possible paths by which noise may be coupled. Both radiated and conducted noise must be considered. Conducted noise lends itself well to circuit and transmission line theory and is simpler to analyze than radiated noise. The latter can accurately be described only by Maxwell's equations. However, it will be seen that circuit theory can sometimes provide valuable insight into the coupling path of radiated noise. We shall use both approaches, comparing them where appropriate.

We must be mindful of the fact that EMI control is an inexact science because of the complexity of practical hardware. It is virtually impossible to define the geometry of a noise source with accuracy sufficient to predict interference within +200 percent or –66 percent, a factor of three. Usually the margin of error is much greater than that. The main purpose of studying EMI is to understand the guidelines that we must follow in the design of an electronic device and to learn how we should *expect* particular designs to affect the coupled noise. If we compare the predicted noise coupling resulting from two different designs, the design that predicts less coupling is preferable from an EMI standpoint. To some extent, we can estimate whether we have a noise problem, but we must never rely on the accuracy normally expected in other typical circuit design problems.

2

METHODS OF EMI ANALYSIS

To understand electromagnetic interference, we begin by looking at how the interference is generated. Of course it is preferable never to generate the interference in the first place, but usually it cannot be avoided. One exception is contact sparking, where the spark serves no useful purpose and even causes damage to the contacts. Adding a diode or capacitor to suppress the spark will not only prolong contact life but also eliminate a possible source of electromagnetic interference. Yet the appliance, controlled by the contacts, functions normally without the spark. Another example is flashover in a high-voltage transformer, where again the sparking is undesirable. Preventive measures for such sparking are well known in electrical engineering and will not be discussed further.

The principal theme of this book will involve applications in which we cannot eliminate the interference at its source. Instead, we must control its path to the affected device. An example is an automotive ignition system. The spark can easily be eliminated from mechanical ignition points, but we cannot eliminate it from the spark plugs without rendering the engine inoperative. The radio-frequency (*RF*) energy generated by the spark is useless but unavoidable. So in this example the energy must somehow be prevented from propagating to any device with which it can interfere. Another example, and by far the most common one, is a digital computer. Here the RF energy does serve a necessary purpose. Without it, the digital logic circuits would have to be very slow, and the speed of the computer would be greatly reduced. However, it is unnecessary and undesirable for the energy to be radiated or conducted outside the computer or to the wrong circuits within the computer. If the RF energy can be confined to its intended circuits within the computer, it serves its purpose without causing interference.

If the RF energy must be intentionally conducted or radiated outside the device, the application is more difficult to treat. For example, the RF energy from a licensed transmitter does not serve its intended purpose until it is radiated into space. Spurious frequencies can and should be eliminated, but the licensed frequency itself may undesirably affect some device. Then the only course of action is to prevent the energy from reaching the device. This is a susceptibility problem, as mentioned in Chapter 1. Appliances containing no high-speed semiconductors, such as toasters, irons, and electric mixers, pick up such energy but are unaffected. Other appliances, such as audio systems or TV receivers, can be severely degraded by RF energy at undesired frequencies.

The above alternatives are summarized in Table 2-1. As we study these techniques throughout this book they will become more meaningful. Since several different remedies are usually possible, we should always remember which is the simplest and least expensive.

Whenever RF energy must be intentionally generated, its path to any adversely affected device must be controlled. To predict its effect, we must determine the amount of attenuation between the energy source and the affected device. We may use circuit theory or field theory, and the relative advantages of each approach will now be qualitatively compared using a simple example. A detailed quantitative analysis follows in Chapter 3.

Consider three parallel wires, as shown in Figure 2-1. Two of the wires are signal leads, labeled 1 and 2. The third is a common signal-return lead, labeled G since it is often connected to ground. This is a typical arrangement in a multi-conductor cable. If one signal lead carries a high-frequency voltage and current, these

TABLE 2-1 Alternatives for
Eliminating EMI in Order of Preference

1. Eliminate source:
   - Suppress contact spark.
   - Prevent high-voltage flashover.
   - Suppress parasitic oscillations.
   - Use slower transition times.

2. Confine within intended circuits:
   - Filter leads not requiring RF.
   - Locate signal leads near ground.
   - Use judicious circuit layout.
   - Shield the wires.
   - Shield the entire source chassis.

3. Control frequencies of emissions.
   - Suppress unnecessary harmonics.
   - Use minimum bandwidth.

4. Protect affected device:
   - Use components not affected by RF.
   - Filter leads not requiing RF.
   - Shield the wires.
   - Shield the entire device.

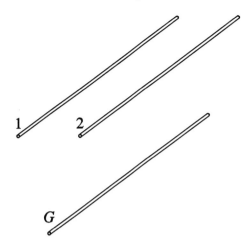

Figure 2-1   Two signal leads and their common ground return

will be coupled into the other signal lead. By analyzing the coupling mechanism, we can predict the attenuation in the coupling path. Ideally, of course, the attenuation would be infinite and there would be no coupling.

2-1 CIRCUIT THEORY

All electrical engineers should be familiar with basic circuit theory. A coupling path may be defined in terms of stray capacitances, inductances, and sometimes resistances. Any two electrical conductors possess a capacitance between them, unless the electric flux between them is somehow diverted or canceled. Also, any practical current path links its own magnetic flux and therefore has a self inductance. Furthermore, any two current paths possess a mutual inductance unless the magnetic flux linking both paths is somehow diverted or canceled.

In the example shown in Figure 2-1, there is some capacitance between each pair of wires and some inductance in each wire. The capacitance and inductance values depend on the wire diameter, the wire spacing, and the relative permittivity and permeability of their insulation. Symbols representing these capacitances and inductances appear in Figure 2-2.

We recall that the capacitance between any two leads is equal to the electric charge on one lead divided by the voltage between them that causes it. The capacitance formula for this simple geometry will be derived in Chapter 3. The inductances, however, require more careful analysis. Although individual inductances appear in each lead in Figure 2-2, no inductance is meaningful by itself. It must form part of a closed current loop. Any two of the three leads in Figure 2-2 form a long, narrow closed loop if we join the two leads at each end of the cable. We will designate the two loops of interest as loops 1 and 2. To form loop 1 we connect lead 1 to lead G at one end of the cable, and do the same at the other end. We form loop 2 similarly, using leads 2 and G. The total inductance in either loop

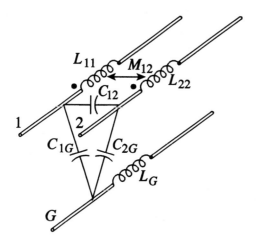

Figure 2-2 Capacitance and inductance in and between leads

is then equal to the magnetic flux linking the loop divided by the current producing the flux.

Like self inductance, mutual inductance is meaningful only with respect to closed current loops. Mutual inductance is defined only between two different current loops, such as loops 1 and 2 above. It is equal to the magnetic flux linking either loop divided by the current in the other loop that generates the flux. Since lead G forms a part of each loop, we cannot define any meaningful mutual inductance between lead G and lead 1 or 2. The only meaningful mutual inductance is between leads 1 and 2, which we show by a dot at one end of each coil in Figure 2-2. We will derive geometric formulas for these self and mutual inductances in Chapter 3. For the present, we must acquire the habit of thinking of the inductance only of a closed circuit and never of a single lead by itself.

With practice, it is possible to locate most or all of the stray inductances and capacitances. Calculating their magnitudes, however, is often quite difficult. Formulas exist[1,2] for simple geometries other than Figure 2-1, but practical cases usually involve many approximations. Sometimes they can be measured with suitable instruments, but doing so requires extensive analysis to consider all possible cases.

Another shortcoming of the circuit approach is that the stray capacitances and inductances must be considered as lumped elements. In reality, they are usually distributed over a region. If all dimensions in the coupling path are small when compared to a quarter wavelength (at every frequency of concern), then the lumped elements provide a good approximation to the distributed parameters. For this to be true, both the source and the affected device must be much smaller than a quarter wavelength. The separation distance between them must also be that small. Circuit

1. John D. Kraus, *Electromagnetics,* 4th ed. (New York, N.Y.: McGraw-Hill Publishing Company, 1992), pp. 235–38.

2. Nannapaneni Narayana Rao, *Elements of Engineering Electromagnetics,* 3rd ed. (Englewood Cliffs, N.J.: Prentice Hall Press, 1991), p. 251.

theory is then a meaningful technique, and we must use field theory only to find the individual inductance and capacitance values. Because of its familiarity to electrical engineers, circuit theory will form an important part of the analyses contained in this text. If the size conditions described above are not satisfied, however, then circuit theory provides little or no information about the coupling path.

2-2 FIELD THEORY

Although basic field theory is a part of the normal electrical engineering curriculum, many non-specialists may have forgotten it. Instead of the capacitance between two conductors, we define the electric field **E** and calculate it as a function of position in space. Similarly, instead of inductance, we define the magnetic field **H**, which also varies with position in space. This approach is much more general and is theoretically applicable to every conceivable geometry.

If we apply a potential difference between lead 1 and lead G of Figure 2-1, an electric field results between the leads, as shown in Figure 2-3. We assume the space between the wires to be a homogeneous dielectric, such as air or a solid mass of insulation with no air spaces. For clarity, the field is shown only at two discrete

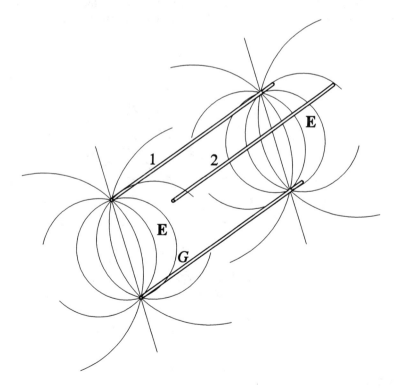

Figure 2-3   Electric field between signal lead and ground return

points on the wires, although it really exists continuously along them. It is evident in Figure 2-3 that some of these electric field lines intersect lead 2. These correspond to C_{12} and C_{2G} of Figure 2-2. The equipotential surfaces are normal to the field lines and parallel with the wires. One such surface, which must pass through lead 2, is at the same potential as that lead. The potential on this surface is the voltage capacitively coupled from lead 1 to lead 2. If we can find that potential, then we know the coupled voltage.

The magnetic field due to a current flowing in lead 2 and returning via lead G (loop 1 defined above) appears in Figure 2-4. Some, but not all, of the magnetic flux lines also link loop 2. This flux corresponds to the sum of the mutual inductance M_{12} and the self inductance L_G. If we can identify the flux lines that link both loops, then we know the inductively coupled voltage.

Both the electric and magnetic fields appear in Figure 2-5. With the homogeneous dielectric assumed for this example, the equipotential surfaces of the electric field coincide with the magnetic field lines. It will be seen that this property causes the capacitive and inductive coupling to vary proportionally as the wire spacing changes. This observation will have important consequences later.

Unfortunately, for most other geometries, the mathematical functions are usually very complicated and cannot be expressed in closed form. Series approximations are then necessary, and the complexity increases with the degree of accuracy required. Digital computers have made such approximations practical, but even

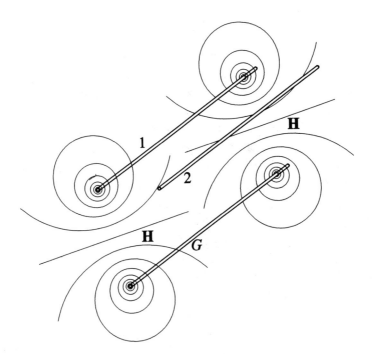

Figure 2-4   Magnetic field between signal lead and ground return

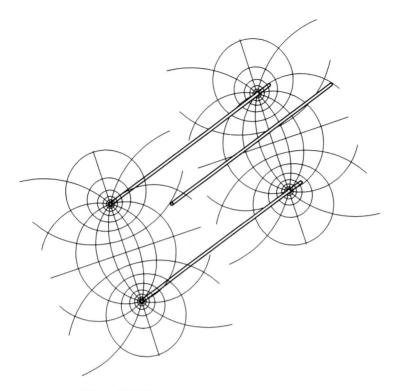

Figure 2-5   Electric and magnetic fields superimposed

these have their limitations. Typical field theory problems might require several days, weeks, or possibly years of processor time unless run on a multi-processor supercomputer. Even then, the processor time may become too expensive. When a solution can be found, however, it will then be applicable to all system sizes and spacings, with no restriction with respect to wavelength. The Finite Element Method[3] is a very powerful tool for this purpose, but it is beyond the scope of this book.

If the separation distances are much *larger* than a quarter wavelength, with empty space between the devices, then the mathematics associated with field theory becomes simpler. The **E** and **H** fields then follow a simple relationship, and a nearly exact solution is often possible. So, for large separation distances, not only is circuit theory useless, but field theory is comparatively simple and is the only applicable method.

We have thus deduced that two different methods of analysis are necessary, depending on the separation distance between the interference source and the affected device. For small separation distances, circuit theory is usable for a quali-

---

3. H. R. Schwarz, *Finite Element Methods* (San Diego, Calif.: Academic Press, Inc., 1984).

tative understanding of the problem. We will explain the electromagnetic coupling paths using circuit theory, describing the physical causes of the stray inductances and capacitances. Then, to quantify these parameters for simple geometries, we will use field theory. Particular emphasis will be placed on understanding the relationship between the fields and their corresponding circuit elements. With an understanding of the coupling phenomena for simple geometries, we will develop a sense for more realistic cases.

For large separation distances, we will use field theory to characterize the coupling path. We must study the method by which RF energy is coupled from the source circuit into free space, and from free space into the affected circuit. This is antenna theory, which may be unfamiliar to some electrical engineers because it is not always a required course of study. Sufficient antenna theory will be presented for an understanding of RF interference coupling to and from free space. This theory will then permit us to find the total attenuation in the coupling path. We will consider these cases beginning in Chapter 8.

2-3 COMPARATIVE ANALYSIS EXAMPLES

If an alternating current (AC) potential V_{1G} exists on lead 1 with respect to lead G in Figure 2-2, then a potential V_{2G} will be capacitively coupled onto lead 2. In the simplest case where the termination impedances on lead 2 are very high, this voltage is simply

$$V_{2G} = \frac{C_{12}V_{1G}}{C_{12} + C_{2G}} \tag{2-1}$$

As previously mentioned, inductances are meaningful only when considered as parts of a closed loop of current. The total inductance of loop 1, defined earlier, is then equal to $L_{11} + L_G$ and the inductance of loop 2 is $L_{22} + L_G$. The mutual inductance of interest here is that between loops 1 and 2. We will show that this mutual inductance is equal to $M_{12} + L_G$.

If an AC voltage source V_1 is in series with loop 1, then a voltage V_2 will be inductively coupled into loop 2. The simplest case here is when the termination impedances in loop 1 are very low. Then it may be shown that the voltage V_2 is

$$V_2 = \frac{M_{12} + L_G}{L_{11} + L_G}V_1 \tag{2-2}$$

If we can calculate, measure, or estimate the capacitances and inductances, then we can find the transfer functions of the $L\text{-}C$ circuit as for any two-terminal-pair network. This is how we calculate the attenuation using circuit theory, which we will demonstrate for more general cases in Chapter 3.

The alternate approach is to consider the electric and magnetic fields around the wires. To analyze them, we must find mathematical functions for the fields that satisfy Maxwell's equations and the boundary conditions. When we know both fields at all points within or near the wires, we can calculate all voltages and currents throughout the circuit. Then we know the attenuation in the coupling path.

There will be no variation of the fields in the direction parallel to the wires if the wire length is short enough. If the wire length is short compared to a quarter wavelength at the frequency of operation, the conductors behave as a lumped capacitance and inductance, which can be treated separately. Field theory will then yield results similar to circuit theory if we properly calculate the capacitance and inductance. On the other hand, if the circuit dimensions do not satisfy the above condition, as when the wires are too long, then field theory is the only approach that yields meaningful results.

Since the solutions to Maxwell's equations are very complex except with simple boundary conditions, we will attempt to keep them as simple as possible. We will treat each inductance and capacitance separately where possible and use circuit theory to interconnect them. Field theory will then serve to evaluate the inductances and capacitances under simpler boundary conditions. This method will be described in Chapter 3.

2-4 PROBLEMS

1. Derive Equations (2-1) and (2-2) in the text.

2. The total capacitance measured between leads 1 and 2 in Figure 2-2 (with lead G floating) is 3 pF. This is not equal to C_{12} because the series combination of C_{1G} and C_{2G} forms part of this measured capacitance. Similarly, the total capacitance measured between leads 1 and G is 4.8 pF, and the total capacitance measured between leads 2 and G is also 4.8 pF. Calculate C_{12} and C_{2G}.

3. Let V_{1G} represent an AC voltage on lead 1 with respect to lead G, and similarly for V_{2G}. Calculate V_{2G}/V_{1G} for the above capacitance values, assuming that all termination impedances are very large and thus can be neglected.

4. With lead 1 shorted to lead 2 at their ends farthest from the observer in Figure 2-2, the inductance of the resulting loop is measured and found to be 1 μH. From circuit theory, this inductance must be equal to $L_{11} + L_{22} - 2M_{12}$. Similarly, the measured inductance of the loop consisting of leads 1 and G is 2 μH, and the measured inductance of the loop consisting of leads 2 and G is also 2 μH. Calculate $L_{11} + L_G$, $L_{22} + L_G$, and $M_{12} + L_G$.

5. Let V_1 represent an AC voltage source in series with loop 1 (defined in the text). Let V_2 be the voltage induced into loop 2 due to inductive coupling. Calculate V_2/V_1 for the above inductance values, assuming that the termination impedances in loop 1 are very small and can be neglected.

6. In Problem 4, you were not asked to calculate L_{11}, L_{22}, M_{12}, or L_G individually.
   (a) Why would this be impossible for a realistic situation?
   (b) Why is it unnecessary?

3

ELECTRIC AND MAGNETIC COUPLING BETWEEN NEARBY DEVICES

Between nearby devices, at frequencies for which their dimensions are much shorter than a quarter wavelength, we may consider electric field coupling separately from magnetic field coupling. To analyze the coupling between two signal wires having a common ground return, we first represent the electric fields as capacitors and the magnetic fields as inductors.

The schematic diagram of a typical circuit using the two signal wires and their common return wire is shown in Figure 3-1. The source and load impedances

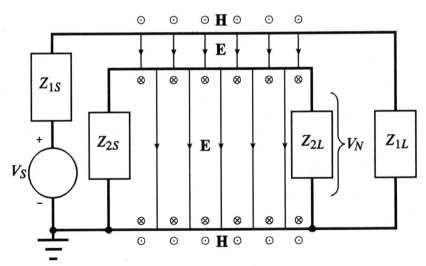

Figure 3-1 Electric and magnetic fields of two signal paths

connected to lead 1 are Z_{1S} and Z_{1L}, respectively, and those connected to lead 2 are Z_{2S} and Z_{2L}. We wish to find the stray "noise" voltage V_N, coupled into lead 2, in proportion to the source voltage V_S driving lead 1. In a real circuit, a voltage source would be connected to lead 2, to generate the signal intended for this lead. To analyze the stray signal by itself, we assume that the desired signal is zero. Therefore, no voltage source appears on lead 2 in Figure 3-1, but its impedance Z_{2S} is still present.

We may approximate the fields in this circuit by lumped elements, as shown in the schematic of Figure 3-2. If all fields are negligible except those shown, then $C_{1G}, C_{12}, C_{2G}, L_{11}, L_{22}, M_{12}$, and L_G can be calculated from the geometry of the leads. We will discuss this procedure later, but will now assume that we know (or can measure) their values.

3-1 CAPACITIVE OR ELECTRIC COUPLING

To consider capacitive (electric-field) coupling, we may assume that L_{11}, L_{22}, M_{12}, and L_G in Figure 3-2 are zero. For simplicity, we will consider only resistive termination impedances, designated R_{1S}, R_{1L}, R_{2S}, and R_{2L}. Let V_N represent the noise voltage appearing across R_{2L} due to V_S. We can then redraw the circuit as shown in Figure 3-3. In this form, the capacitances C_{1G}, C_{12}, and C_{2G} are easy to visualize. However, they cannot be measured or calculated individually. The capacitance between lead 1 and ground, for example, is *not* C_{1G} alone but the parallel combination of C_{1G} with C_{12} and C_{2G} in series. If we write the loop or node equations in terms of these capacitances, the expression for V_N will later become unnecessarily complicated. We could then simplify it (see Problems 1 through 4 at

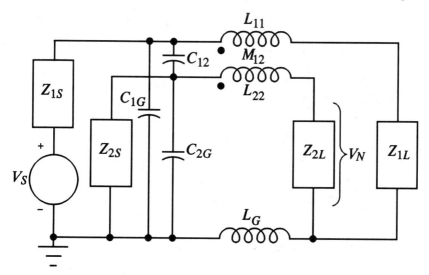

Figure 3-2 Capacitances and inductances in two signal paths

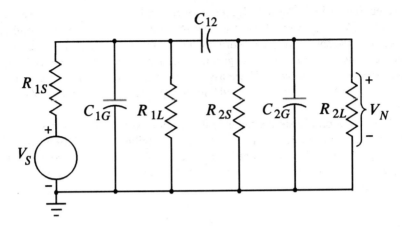

Figure 3-3 Capacitances in two signal paths as a delta network

the end of this chapter), but it is better to avoid the complicated expression alto-gether. We do so by substituting the wye (star) equivalent of the delta circuit of C_{1G}, C_{12}, and C_{2G}. The resulting circuit appears in Figure 3-4. We could express the new capacitances C_1, C_2, and C_G in terms of the delta capacitances of Figure 3-3. However, this is unnecessary since it will be easier to calculate or measure C_1, C_2, and C_G directly. We also will discover that these capacitances are closely related to the inductive parameters to be discussed in the next section.

A wye circuit is most conveniently analyzed via its loop equations. Therefore, the expressions become even simpler if we express each capacitance as an *elast-ance*[1], which is the reciprocal of capacitance and is designated S. The unit of

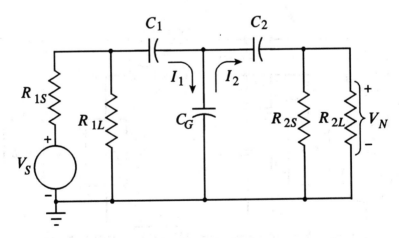

Figure 3-4 Capacitances in two signal paths as a wye network

1. Donald G. Fink and Donald Christensen, *Electronic Engineer's Handbook,* 3rd ed. (New York, N.Y.: McGraw-Hill Publishing Company, 1989), Chapter 1, p. 14.

elastance used in older literature is the *daraf*, but since this is not a standard (SI) unit we will simply use F^{-1} or, more practically, μF^{-1} or pF^{-1}. A capacitance of 2 pF corresponds to an elastance of 0.5 pF^{-1}. Thus, $S_1 = 1/C_1$, $S_2 = 1/C_2$, and $S_G = 1/C_G$. We shall use these quantities in lieu of the respective capacitances for the remainder of this section. Later, when we express capacitances and inductances in terms of physical dimensions, we will note a close resemblance between inductance and elastance formulae. This will produce a convenient analogy between electric and magnetic coupling if we use elastance instead of capacitance, and use the wye circuit instead of the delta.

Let R_{1P} represent the parallel combination of R_{1S} and R_{1L}, and similarly for R_{2P}. Then, by using the Thévenin equivalent of the combination of V_S, R_{1S}, and R_{1L}, we reduce the circuit to two loops, designated 1 and 2, and with loop equations as follows:

$$R_{1P}I_1 + \frac{S_1I_1}{j\omega} + \frac{S_G(I_1 - I_2)}{j\omega} = \frac{R_{1L}}{R_{1S} + R_{1L}}V_S$$

$$\left(R_{1P} + \frac{S_1 + S_G}{j\omega}\right)I_1 - \frac{S_G}{j\omega}I_2 = \frac{R_{1P}}{R_{1S}}V_S \tag{3-1}$$

$$\frac{S_G(I_2 - I_1)}{j\omega} + \frac{S_2I_2}{j\omega} + R_{2P}I_2 = 0$$

$$-\frac{S_G}{j\omega}I_1 + \left(R_{2P} + \frac{S_2 + S_G}{j\omega}\right)I_2 = 0 \tag{3-2}$$

Solving Equations (3-1) and (3-2) for I_2 then yields

$$I_2 = \frac{\dfrac{R_1P}{R_{1S}} \cdot \dfrac{S_G}{j\omega}V_S}{\left(R_{1P} + \dfrac{S_1 + S_G}{j\omega}\right)\left(R_{2P} + \dfrac{S_2 + S_G}{j\omega}\right) - \left(\dfrac{S_G}{j\omega}\right)^2} \tag{3-3}$$

Then, since $V_N = I_2R_{2P}$,

$$V_N = \frac{\dfrac{R_{1P}R_{2P}}{R_{1S}} \cdot \dfrac{S_G}{j\omega}V_S}{\left(R_{1P} + \dfrac{S_1 + S_G}{j\omega}\right)\left(R_{2P} + \dfrac{S_2 + S_G}{j\omega}\right) - \left(\dfrac{S_G}{j\omega}\right)^2}$$

$$= \frac{j\omega S_G R_{1P}R_{2P}/R_{1S}}{(j\omega R_{1P} + S_1 + S_G)(j\omega R_{2P} + S_2 + S_G) - S_G^2}V_S$$

$$= \frac{j\omega S_G R_{1P}R_{2P}/R_{1S}}{(j\omega)^2 R_{1P}R_{2P} + j\omega[R_{1P}(S_2 + S_G) + R_{2P}(S_1 + S_G)] + (S_1 + S_G)(S_2 + S_G) - S_G^2}V_S \tag{3-4}$$

The elastances S_1 and S_2 appear only in expressions in which they are added to S_G. This is physically meaningful since the measured capacitance from lead 1 to ground is $1/(S_1 + S_G)$, and similarly for lead 2. We also will see later that the geometric formulas for $S_1 + S_G$ and $S_2 + S_G$ are simpler than those for the individual capacitances. Therefore, we will keep the elastances in this form.

Equation (3-4) is easier to understand if we separate the denominator into its factors, as follows:

$$V_N + \frac{S_G V_S}{R_{1S}} \cdot \frac{j\omega}{(j\omega + s_a)(j\omega + s_b)} \tag{3-5}$$

where

$$s_a, s_b = \frac{S_1 + S_G}{2R_{1P}} + \frac{S_2 + S_G}{2R_{2P}} \pm \sqrt{\left(\frac{S_1 + S_G}{2R_{1P}} - \frac{S_2 + S_G}{2R_{2P}}\right)^2 + \frac{S_G^2}{R_{1P}R_{2P}}} \tag{3-6}$$

Note that since the radicand in Equation (3-6) cannot be negative, s_a and s_b must be real. They are the poles of the transfer function, which also has a zero at the origin.

At low frequencies, the $j\omega$ terms in the denominator become insignificant, and Equation (3-5) reduces to

$$V_N = \frac{S_G V_S}{R_{1S}} \cdot \frac{j\omega}{s_a s_b} = j\omega \cdot \frac{S_G R_{1P} R_{2P} V_S}{R_{1S}[(S_1 + S_G)(S_2 + S_G) - S_G^2]} \tag{3-7}$$

Thus, at low frequencies, the resistances become insignificant, the capacitances behave as a current divider, and I_2 and V_N are proportional to frequency.

At high frequencies, the s_a and s_b terms in the denominator of Equation (3-5) become insignificant, and the equation reduces to

$$V_N = \frac{S_G V_S}{R_{1S}} \cdot \frac{j\omega}{(j\omega)^2} = \frac{S_G V_S}{j\omega R_{1S}} \tag{3-8}$$

At these frequencies, all capacitive reactances are so small that only C_G has any effect since there is no resistance in series with it. As the frequency increases, C_G shunts more current to ground, and V_N decreases in inverse proportion to frequency.

The mid-range behavior of Equation (3-5) becomes important if s_a and s_b are far apart, which occurs if the square root term in Equation (3-6) is significant. It is evident from Equation (3-6) that there are two conditions that can cause this to occur. One possibility is that $(S_1 + S_G)/R_{1P}$ differs considerably from $(S_2 + S_G)/R_{2P}$. Then, for the middle frequency range, the resistance dominates in one loop while the capacitive reactance dominates in the other. If R_{2P} dominates, then the middle frequency range occurs where R_{2P} behaves as an open circuit and R_{1P} behaves as a short circuit. Then S_1 forms a voltage divider with S_G. If R_{1P} dominates, then it behaves as an open circuit, R_{2P} behaves as a short circuit, and S_2 forms a current divider with S_G. In either case, V_N is nearly independent of frequency over this range. The other possibility is that S_G is much larger than S_1 and S_2. Then, for the middle frequency range, S_G appears as an open circuit while S_1 and S_2 appear as short circuits. Then R_{1P} and R_{2P} form a voltage divider, and again V_N is nearly constant over this frequency range. In all these cases, the frequency range under consideration is that for which ω dominates over s_a but is insignificant compared to s_b. Then Equation (3-5) reduces to

$$V_N = \frac{S_G V_S}{R_{1S} s_b} \tag{3-9}$$

where s_b is as given in Equation (3-6). This confirms that for this frequency range, V_N is constant, independent of frequency. This value of V_N is its maximum value over all frequencies. If s_a is near to s_b, this middle frequency range is small. Even then, it is true that V_N can never exceed the value given in Equation (3-9).

For example, assume $R_{1P} = R_{2P} = 2000\ \Omega$, $S_G = 0.2083\ \text{pF}^{-1}$ (or $C_G = 4.8\ \text{pF}$), and $S_1 + S_G = S_2 + S_G = 0.25\ \text{pF}^{-1}$, which are typical values. Then from Equation (3-6) $s_a = 20.85 \cdot 10^6$ rad/s or 3.32 MHz, and $s_b = 229.15 \cdot 10^6$ rad/s or 36.47 MHz. If $R_{1L} \gg R_{1S}$ so that $R_{1P} \approx R_{1S}$, Equation (3-4) or (3-5) states that at $f = 10$ MHz (where $\omega = 2\pi f$), $|V_N| = 0.416\ |V_S|$. Since $s_a < 10$ MHz $< s_b$, Equation (3-9) should give a reasonable approximation for $|V_N|$, and it does, since it says that $|V_N| \approx 0.455\ |V_S|$.

A graph of the ratio $|V_N/V_S|$ versus frequency appears in Figure 3-5 for the R and S values given above. The straight-line approximations of the curve for $C_G = 4.8$ pF are superimposed on the graph, which is then called a Bode plot. The intersections of these lines occur at the frequencies s_a and s_b, sometimes known as the "corner" frequencies. Curves are also shown for other values of C_G, with $S_1 + S_G$ and $S_2 + S_G$ still 0.25 pF^{-1}.

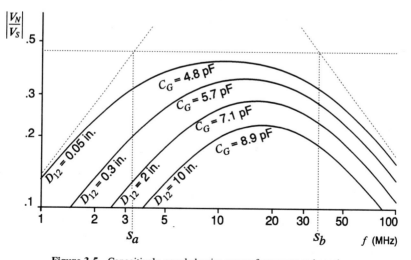

Figure 3-5   Capacitively-coupled noise versus frequency and spacing

We can plot a more general graph of V_N versus f if we define four dimensionless parameters as follows:

$$P_f = \omega \sqrt{\frac{R_{1P}R_{2P}}{(S_1 + S_G)(S_2 + S_G)}} \qquad P_G = \frac{S_G}{\sqrt{(S_1 + S_G)(S_2 + S_G)}}$$

$$P_D = \frac{R_{1P}(S_2 + S_G) + R_{2P}(S_1 + S_G)}{2\sqrt{R_{1P}R_{2P}(S_1 + S_G)(S_2 + S_G)}} \qquad P_V = \frac{V_N R_{1S}}{V_S \sqrt{R_{1P}R_{2P}}}$$

$$(3\text{-}10)$$

Equation (3-4), in terms of these parameters, then becomes

$$P_V = \frac{jP_f P_G}{(jP_f)^2 + j2P_f P_D + 1 - P_G^2} \tag{3-11}$$

Whenever the loops are identical, that is, whenever $R_{1P} = R_{2P}$ and $S_1 + S_G = S_2 + S_G$ as in the above example, $P_D = 1$. Figure 3-6 shows a plot of P_V versus P_f, for $P_D = 1$, and for P_G calculated from the values of C_G in the above example. The corner frequency points, also appearing in the figure for $P_G = 0.833$, are

$$P_{fa}, P_{fb} = P_D \pm \sqrt{P_D^2 - 1 + \tfrac{2}{G}} \tag{3-12}$$

or, since $P_D = 1$,

$$P_{fa}, P_{fb} = 1 \pm \sqrt{1 - 1 + P_G^2} = 1 \pm P_G \tag{3-13}$$

We can use this graph for any geometry for which $P_D = 1$. For $R_{1P} = R_{2P} = 2000\ \Omega$, and for $S_1 + S_G = S_2 + S_G = 0.25\ \text{pF}^{-1}$, from the example,

$$P_f = \omega \sqrt{\frac{2000\ \Omega \cdot 2000\ \Omega}{0.25\ \text{pF}^{-1} \cdot 0.25\ \text{pF}^{-1}}} = \omega \cdot 8\ \text{ns} = 0.0503 \cdot f \tag{3-14}$$

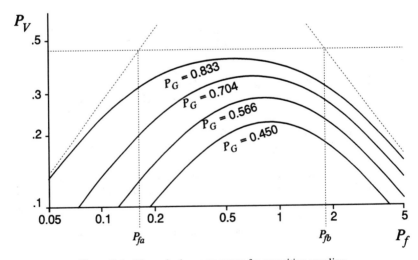

Figure 3-6 Dimensionless parameters for capacitive coupling

with f in megahertz. This verifies the correspondence between the scales of Figures 3-5 and 3-6.

For other values of R and C, we first calculate P_D. If $P_D = 1$, we can use Figure 3-6; otherwise, we must plot a different graph. If $P_D = 1$, we calculate P_G to find which curve in Figure 3-6 is nearest to it. Then we calculate the coefficient of f in P_f to scale the figure. For example, if R_{1P} and R_{2P} are 400 Ω instead of 2000 Ω but the capacitances remain the same, then for the same frequency, P_f decreases by a factor of five. The effect is to increase the frequency scale of Figure 3-6 by this factor. Thus, for lower termination resistances, the frequency range of maximum concern ($s_a < 2\pi f < s_b$) moves upward. As this frequency range becomes greater than 1000 MHz, the wire lengths are no longer short compared to a quarter wavelength. Then the analysis is no longer applicable. Capacitive coupling is therefore of greatest concern for high termination impedances.

In the earlier example, we also assumed that $R_{1L} \gg R_{1S}$. If this is not true (but still $R_{1P} = R_{2P}$), then the factor of $R_{1S}/\sqrt{R_{1P}R_{2P}}$ is no longer approximately equal to one. This changes the voltage scale in Figure 3-6 by the factor of $R_{1S}/\sqrt{R_{1P}R_{2P}}$, according to the definition of P_V.

Capacitively coupled noise is often modeled by a current source. The noise current may be injected into lead 2 directly, or it may be applied at the junction of C_2 and C_G in Figure 3-4. If it is injected directly into lead 2, as shown in Figure 3-7, then its magnitude I_N is

$$I_N = V_N \cdot \frac{R_{2P} + (S_2 + S_G)/j\omega}{R_{2P}(S_2 + S_G)/j\omega}$$

$$= I_2 \cdot \frac{j\omega R_{2P} + S_2 + S_G}{S_2 + S_G}$$

$$= \frac{j\omega S_G R_{1P}}{(S_2 + S_G)R_{1S}} \cdot \frac{j\omega R_{2P} + S_2 + S_G}{(j\omega R_{1P} + S_1 + S_G)(j\omega R_{2P} + S_2 + S_G) - S_G^2} V_S \qquad (3\text{-}15)$$

If we instead apply the noise current to the junction of C_2 and C_G, then it becomes equal to I_1 in Figure 3-4, which may then be redrawn as in Figure 3-8. We obtain its magnitude by solving Equations (3-1) and (3-2) for I_1, which is then equal to

$$I_1 = \frac{V_S \cdot \dfrac{R_{1P}}{R_{1S}}\left(R_{2P} + \dfrac{S_2 + S_G}{j\omega}\right)}{\left(R_{1P} + \dfrac{S_1 + S_G}{j\omega}\right)\left(R_{2P} + \dfrac{S_2 + S_G}{j\omega}\right) - \left(\dfrac{S_G}{j\omega}\right)^2}$$

$$= \frac{j\omega R_{1P}}{R_{1S}} \cdot \frac{j\omega R_{2P} + S_2 + S_G}{(j\omega R_{1P} + S_1 + S_G)(j\omega R_{2P} + S_2 + S_G) - S_G^2} V_S \qquad (3\text{-}16)$$

As we can see by comparing Equations (3-15) and (3-16), I_N of Figure 3-7 differs from I_1 of Figure 3-8 only by a factor of $S_G/(S_2 + S_G)$.

From Equation (3-4) we see that we can reduce $|V_N/V_S|$ by increasing S_1 or S_2 or reducing S_G. It is obvious that either capacitance can be reduced (and the

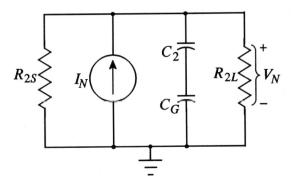

Figure 3-7 Possible noise model

corresponding elastance increased) by increasing the spacing between the respective wires, and conversely. To evaluate this effect quantitatively, however, we must know the values for $S_1 + S_G$, $S_2 + S_G$, and S_G as functions of separation distances. Determination of these capacitances requires field theory.

If we apply a direct-current (DC) voltage V_1 between lead 1 and ground, a charge $+Q$ will accumulate on lead 1. An opposite charge $-Q$ will accumulate on the ground lead. The total capacitance between lead 1 and ground is, by definition, Q/V_1. It is equal to $1/(S_1 + S_G)$ if lead 2 is left floating. Similarly, the total capacitance between lead 2 and ground is $1/(S_2 + S_G)$. The total capacitance between lead 1 and lead 2 is $1/(S_1 + S_2)$. We will use these capacitances to compute S_G itself. The problem that remains is to find these capacitances in terms of the respective separation distances. The values will be only approximate, since the geometry will be idealized and not exact. They will, however, suggest how we can reduce the capacitive coupling.

Consider two parallel wires of radius r, surrounded by homogeneous insulating material. They are displaced by the distances $+s$ and $-s$, respectively, from the z axis of a Cartesian coordinate system. Each of these wires may be lead 1, lead 2, or the ground lead discussed above. The cross section appears in Figure 3-9, with

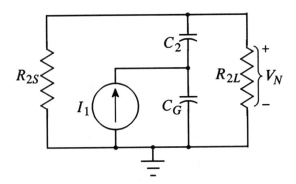

Figure 3-8 Alternative model

the wires labeled a and b. We will now find the capacitance C_{ab} between these wires. We will first show that if a potential difference is introduced between the wires, then the potential V in the space around the wires varies according to the following function:

$$V(x, y) = V_0 \ln \left[\frac{(x + \sqrt{s^2 - r^2})^2 + y^2}{(x - \sqrt{s^2 - r^2})^2 + y^2} \right] \tag{3-17}$$

where V_0 is a scale factor that depends on the potential difference between the wires. The potential V must satisfy Poisson's equation:

$$\nabla^2 V = -\frac{\rho}{\epsilon} \tag{3-18}$$

where ϵ is the permittivity and ρ is the charge density, which can be nonzero only within the conductors. The charge distributes itself on the conductor surfaces such that each surface attains a constant potential. The required charge density ρ is not easy to calculate, and we need not know its value. Instead, we solve the equation by setting ρ equal to zero in the space around the wires. We then impose boundary conditions consisting of equipotential surfaces at the surfaces of the wires. With $\rho = 0$, Poisson's equation becomes Laplace's equation, which is

$$\nabla^2 V = \frac{\partial^2 V}{\partial x^2} + \frac{\partial^2 V}{\partial y^2} + \frac{\partial^2 V}{\partial z^2} = 0 \tag{3-19}$$

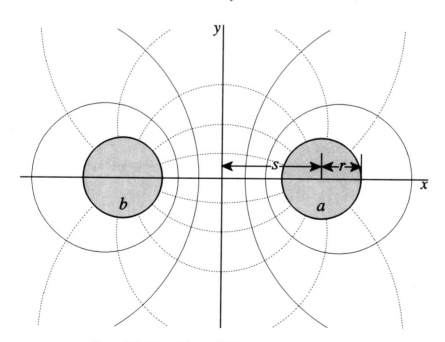

Figure 3-9   Map of electric field between two parallel wires

The required derivatives of the potential function V are

$$\frac{\partial V}{\partial x} = V_0 \left[\frac{2(x + \sqrt{s^2 - r^2})}{(x + \sqrt{s^2 - r^2})^2 + y^2} - \frac{2(x - \sqrt{s^2 - r^2})}{(x - \sqrt{s^2 - r^2})^2 + y^2} \right] \tag{3-20}$$

$$\frac{\partial^2 V}{\partial x^2} =$$

$$V_0 \left\{ \frac{2[(x+\sqrt{s^2-r^2})^2+y^2]-4(x+\sqrt{s^2-r^2})^2}{[(x+\sqrt{s^2-r^2})^2+y^2]^2} - \frac{2[(x-\sqrt{s^2-r^2})^2+y^2]-4(x-\sqrt{s^2-r^2})^2}{[(x-\sqrt{s^2-r^2})^2+y^2]^2} \right\}$$

$$= V_0 \left\{ \frac{-2(x+\sqrt{s^2-r^2})^2+2y^2}{[(x+\sqrt{s^2-r^2})^2+y^2]^2} - \frac{-2(x-\sqrt{s^2-r^2})^2+2y^2}{[(x-\sqrt{s^2-r^2})^2+y^2]^2} \right\} \tag{3-21}$$

and similarly,

$$\frac{\partial V}{\partial y} = V_0 \left[\frac{2y}{(x + \sqrt{s^2 - r^2})^2 + y^2} - \frac{2y}{(x - \sqrt{s^2 - r^2})^2 + y^2} \right] \tag{3-22}$$

$$\frac{\partial^2 V}{\partial y^2} = V_0 \left\{ \frac{2[(x+\sqrt{s^2-r^2})^2+y^2]-4y^2}{[(x+\sqrt{s^2-r^2})^2+y^2]^2} - \frac{2[(x-\sqrt{s^2-r^2})^2+y^2]-4y^2}{[(x-\sqrt{s^2-r^2})^2+y^2]^2} \right\}$$

$$= V_0 \left\{ \frac{2(x+\sqrt{s^2-r^2})^2-2y^2}{[(x+\sqrt{s^2-r^2})^2+y^2]^2} - \frac{2(x-\sqrt{s^2-r^2})^2-2y^2}{[(x-\sqrt{s^2-r^2})^2+y^2]^2} \right\} \tag{3-23}$$

Since there is no variation in the z direction, $\partial^2 v / \partial z^2 = 0$. Thus,

$$\nabla^2 V = \frac{\partial^2 V}{\partial x^2} + \frac{\partial^2 V}{\partial y^2} + \frac{\partial^2 V}{\partial z^2} = 0 \tag{3-24}$$

and Laplace's equation is satisfied, as required.

In addition, we must show that Equation (3-17) satisfies the boundary conditions. At the surface of conductor a, y depends on x as follows:

$$y^2 = r^2 - (x - s)^2 \tag{3-25}$$

The potential V_a at the conductor surface is therefore:

$$V_a = V_0 \ln \left[\frac{(x + \sqrt{s^2 - r^2})^2 + r^2 - (x - s)^2}{(x - \sqrt{s^2 - r^2})^2 + r^2 - (x - s)^2} \right]$$

$$= V_0 \ln \left[\frac{(x^2 + 2x\sqrt{s^2 - r^2} + s^2 - r^2) + r^2 - (x^2 - 2xs + s^2)}{(x^2 - 2x\sqrt{s^2 - r^2} + s^2 - r^2) + r^2 - (x^2 - 2xs + s^2)} \right]$$

$$= V_0 \ln \left(\frac{2x\sqrt{s^2 - r^2} + 2xs}{-2x\sqrt{s^2 - r^2} + 2xs} \right) \tag{3-26}$$

$$= V_0 \ln \left[\frac{(\sqrt{s^2 - r^2} + s)^2}{-(s^2 - r^2) + s^2} \right]$$

$$= 2V_0 \ln \left[\sqrt{\left(\frac{s}{r}\right)^2 - 1} + \frac{s}{r} \right]$$

$$= 2V_0 \cosh^{-1} \left(\frac{s}{r}\right) \qquad \text{(3-26)} \\ \text{(cont'd)}$$

which is a constant potential, independent of x, as it must be on the surface of a perfect conductor. Similarly, at the surface of conductor b,

$$y^2 = r^2 - (x + s)^2 \qquad \text{(3-27)}$$

and the potential V_b on that surface is therefore

$$V_b = V_0 \ln \left(\frac{-2x\sqrt{s^2 - r^2} + 2xs}{2x\sqrt{s^2 - r^2} + 2xs} \right)$$

$$= V_0 \ln \left[\frac{-(s^2 - r^2) + s^2}{(\sqrt{s^2 - r^2} + s)^2} \right]$$

$$= -2V_0 \ln \left[\sqrt{\left(\frac{s}{r}\right)^2 - 1} + \frac{s}{r} \right]$$

$$= -2V_0 \cosh^{-1} \left(\frac{s}{r}\right) \qquad \text{(3-28)}$$

The potential difference, V_{ab}, between the wires, is

$$V_{ab} = V_a - V_b = 4V_0 \cosh^{-1} \left(\frac{s}{r}\right) \qquad \text{(3-29)}$$

If we choose the constant V_0 so that V_{ab} is equal to the actual potential difference between the wires, then Equation (3-17) satisfies the boundary conditions, as required.

To find C_{ab}, we must know the total charge Q on the positive wire. By Gauss's law, it is equal to the total electric flux, Ψ, between the wires. This is equal to the integral of the flux density, D, over any surface enclosing the positive wire. We choose a cylindrical surface having a D-shaped cross section. This surface consists of a section of the yz plane and an infinitely large semicircular cylindrical surface connecting its two edges that are parallel to the wires. The surface integral of D over the semicircular cylindrical surface is zero, since $|D|$ decreases inversely with the square of the radius. Thus the only nonzero part of the integral is on the yz plane, for which

$$Q = \Psi = \oint_s D \cdot ds = -\int_{-\infty}^{\infty} D_x \Big|_{x=0} l\, dy = \epsilon l \int_{-\infty}^{\infty} \frac{\partial V}{\partial x} \Big|_{x=0} dy \qquad \text{(3-30)}$$

where l is the length of the line and ϵ is the permittivity of the insulation surrounding the conductors. If ϵ_r is the *relative* permittivity, or dielectric constant, of this insulation, then

$$\epsilon = \epsilon_r \epsilon_0 \approx \epsilon_r \cdot 8.84 \text{ pF/m} \tag{3-31}$$

Substituting for $\partial V/\partial x$ from Equation (3-20) yields

$$Q = \epsilon l \int_{-\infty}^{\infty} V_0 \left[\frac{2(0 + \sqrt{s^2 - r^2})}{(0 + \sqrt{s^2 - r^2})^2 + y^2} - \frac{2(0 - \sqrt{s^2 - r^2})}{(0 - \sqrt{s^2 - r^2})^2 + y^2} \right] dy$$

$$= 2\epsilon l V_0 \left[\tan^{-1} \frac{y}{\sqrt{s^2 - r^2}} - \tan^{-1} \frac{y}{-\sqrt{s^2 - r^2}} \right]_{-\infty}^{\infty}$$

$$= 4\epsilon l V_0 \left[\tan^{-1} \frac{y}{\sqrt{s^2 - r^2}} \right]_{-\infty}^{\infty} = 4\epsilon l V_0 \left(\frac{\pi}{2} + \frac{\pi}{2} \right) = 4\pi\epsilon l V_0 \tag{3-32}$$

The capacitance C_{ab} is then equal to the charge divided by the potential difference, or

$$C_{ab} = \frac{Q}{V_{ab}} = \frac{4\pi\epsilon l V_0}{4V_0 \cosh^{-1}\left(\frac{s}{r}\right)} = \frac{\pi\epsilon l}{\cosh^{-1}\left(\frac{s}{r}\right)} \tag{3-33}$$

More commonly, we know the diameter d and the spacing D_{ab} of the wires instead of the dimensions r and s used in the above derivation. Since $d = 2r$ and $D_{ab} = 2s$, the formula can be conveniently rewritten as

$$C_{ab} = \frac{\pi\epsilon l}{\cosh^{-1}\left(\dfrac{D_{ab}}{d}\right)} \tag{3-34}$$

Consider, for example, two identical wires in free space with their surfaces separated by a distance equal to their diameter d. Their center-to-center spacing is $2d$. The ratio D_{ab}/d is therefore equal to 2, and from Equation (3-34) the capacitance is approximately 21 pF per meter of wire length. For cases where the separation distance D_{ab} is much larger than the diameter d, the capacitance may be approximated by the simpler formula below:

$$C_{ab} \approx \frac{\pi\epsilon l}{\ln\left(\dfrac{2D_{ab}}{d}\right)} \quad \text{for } D_{ab} \gg d \tag{3-35}$$

For two wires in free space separated by a distance equal to five times their diameter, for example, the capacitance calculated using Equation (3-34) is 12.12 pF/meter. If the capacitance is calculated using Equation (3-35) it is 12.06 pF/meter. Two #30 AWG wires ($d = 0.01$ in.) separated by 5 in. would have $D_{ab}/d = 500$, and from Equation (3-35), $C_{ab} = 4$ pF/meter. For a 1-m length of these two wires, $S_{ab} = 1/C_{ab} = 0.25$ pF^{-1}, which is the value assumed earlier in Figure 3-5. We have thus demonstrated that Figure 3-5 assumes the distances D_{1G} and D_{2G} to be 5 in., using 1-m lengths of #30 AWG wire. The values shown for D_{12} will be justified shortly.

If the dielectric is non-homogeneous, such as partly air and partly plastic insulation, an exact solution is much more difficult. Then we may establish bounds for the actual capacitance by calculating it twice, once with each value of ϵ. The actual capacitance will then fall between the two values.

We can now write the formula for the elastance between leads 1 and 2. We assign lead 1 to lead a and lead 2 to lead b, set D_{ab} equal to D_{12}, and invert Equations (3-34) and (3-35). This elastance is equal to $S_1 + S_2$, and therefore

$$S_1 + S_2 = \frac{\cosh^{-1}\left(\dfrac{D_{12}}{d}\right)}{\pi\epsilon l} \approx \frac{\ln\left(\dfrac{2D_{12}}{d}\right)}{\pi\epsilon l} \quad \text{for } D_{12} \gg d \qquad (3\text{-}36)$$

If the common return wire is another lead of the same diameter d, then elastances $S_1 + S_G$ and $S_2 + S_G$ are determined in the same manner, with the leads and spacings appropriately assigned.

$$S_1 + S_G = \frac{\cosh^{-1}\left(\dfrac{D_{1G}}{d}\right)}{\pi\epsilon l} \approx \frac{\ln\left(\dfrac{2D_{1G}}{d}\right)}{\pi\epsilon l} \quad \text{for } D_{1G} \gg d \qquad (3\text{-}37)$$

$$S_2 + S_G = \frac{\cosh^{-1}\left(\dfrac{D_{2G}}{d}\right)}{\pi\epsilon l} \approx \frac{\ln\left(\dfrac{2D_{2G}}{d}\right)}{\pi\epsilon l} \quad \text{for } D_{2G} \gg d \qquad (3\text{-}38)$$

If, however, the return is a large ground plane, then the fields behave as if an image of the lead were at an equal distance below the plane. Then, for the same charge Q on the lead, only half the potential difference appears between the lead and the ground plane. The capacitance between the lead and the plane is therefore double what it would be between the lead and its image. The corresponding elastance is half. The height h of the lead above the ground plane is also half the distance D_{ab} between the lead and its image. So, if lead 1 is at a height h_1 above a ground plane, then

$$S_1 + S_G = \frac{\cosh^{-1}\left(\dfrac{2h_1}{d}\right)}{2\pi\epsilon l} \approx \frac{\ln\left(\dfrac{4h_1}{d}\right)}{2\pi\epsilon l} \quad \text{for } h_1 \gg d \qquad (3\text{-}39)$$

A similar expression applies to lead 2. Consider a 1-m length of wire separated from a ground plane by a distance equal to twice its diameter. Its center is at a height of $5d/2$ above the ground plane. The elastance calculated using Equation (3-39) is approximately 0.029 pF^{-1} for the 1-m length.

We derived the capacitance formulas assuming two conductors in a homogeneous dielectric, with no other nearby conductors. If we introduce any additional finite-sized conductors, they change the boundary conditions by adding new equipotential surfaces. Then the potential function is no longer that given by Equation (3-17). Of course lead 1 and lead 2, with a ground return of some type, comprise more than two leads. Thus, the above capacitance formulas cannot strictly be used. We could find a more complex solution subject to the new boundary conditions. However, this is hardly warranted since the geometry is probably not exact anyhow. If, however, all conductor diameters are much smaller than their spacing, the potential is then already nearly constant along the whole surface of each conductor. Any new conductors parallel to the z axis do not then affect the potential or field, and the expression of Equation (3-17) still applies. Therefore, to simplify the solution, we will hereafter assume that $d \ll D_{1G}$, $d \ll D_{2G}$, and $d \ll D_{12}$, or equivalently, $r \ll s$ in all three cases.

Although S_G could not be calculated directly, it may now be determined easily using the derived expressions for $S_1 + S_G$, $S_2 + S_G$, and $S_1 + S_2$:

$$S_G = \frac{(S_1 + S_G) + (S_2 + S_G) - (S_1 + S_2)}{2}$$

$$= \frac{\ln\left(\dfrac{2D_{1G}}{d}\right) + \ln\left(\dfrac{2D_{2G}}{d}\right) - \ln\left(\dfrac{2D_{12}}{d}\right)}{2\pi\epsilon l}$$

$$= \frac{\ln\left(\dfrac{2D_{1G}D_{2G}}{dD_{12}}\right)}{2\pi\epsilon l} \tag{3-40}$$

This expression can now be used to verify the correspondence between S_G and D_{12} in Figure 3-5, recalling that we assumed that $D_{1G} = D_{2G} = 5$ in. and $d = 0.01$ in.

By substituting Equations (3-37), (3-38), and (3-40) into Equation (3-4), we obtain the expression for V_N in terms of D_{1G}, D_{12}, and D_{2G}:

$$V_N = \frac{j\omega 2\pi\epsilon l \dfrac{R_{1P}R_{2P}}{R_{1S}} \cdot \ln \dfrac{2D_{1G}D_{2G}}{dD_{12}}}{4\left(j\omega\pi\epsilon l R_{1P} + \ln\dfrac{2D_{1G}}{d}\right)\left(j\omega\pi\epsilon l R_{2P} + \ln\dfrac{2D_{2G}}{d}\right) - \left(\ln\dfrac{2D_{1G}D_{2G}}{dD_{12}}\right)^2} V_S \tag{3-41}$$

As an example, consider two #30 AWG wires, designated 1 and 2, in free space, each wire 1 m long. Let the wires be separated by 2 in., with a common return lead, also #30 AWG, separated from each signal lead by 5 in. This is the geometry assumed for Figure 3-5, so $S_1 + S_G = S_2 + S_G = 0.25$ pF^{-1} and $C_G = 7.1$ pF ($S_G = 0.141$ pF^{-1}). Let lead 1 be driven by a 5-V, 50-MHz sine wave signal, and let all source and load impedances be 1000 Ω each. The resistances R_{1P} and R_{2P} of their parallel combinations are each then 500 Ω. The wavelength is approximately 6 m, so the wire lengths are less than a quarter wavelength, but not by much. Nevertheless, we neglect inductance in this analysis. This may introduce some error, which we shall ignore.

Since R_{1P} and R_{2P} are 2000 Ω for the graph in Figure 3-5, we cannot directly use Figure 3-5 for this example. We note that since the two loops are identical, $P_D = 1$, so we can use Figure 3-6 by calculating P_G and the coefficients of P_f and P_V. From the capacitances mentioned above, we find that $P_G = 0.566$. Since $f = 50$ MHz, $P_f = 2\pi \cdot (50 \cdot 10^6) \cdot 500/(0.25 \cdot 10^{12}) = 0.3142$. Thus, we can read P_V from the $P_G = 0.566$ curve in Figure 3-6, at $P_f = 0.3142$. Its value is 0.25, and from this we calculate $V_N = 0.25 \cdot (5 \text{ volts}) \cdot 500 \ \Omega \ / \ 1000 \ \Omega = 0.625$ volts. As the frequency increases, so does V_N until its peak is reached, after which V_N decreases as shown.

If the return path is a ground plane then we must assume images of leads 1 and 2 below the ground plane. Then we use Equation (3-39) to calculate $S_1 + S_G$ and $S_2 + S_G$. The resulting graph will still resemble Figure 3-5.

3-2 INDUCTIVE OR MAGNETIC COUPLING

To analyze the inductive (magnetic-field) coupling we return to Figure 3-2 and now assume that C_{1G}, C_{12} and C_{2G} are zero. Again, for simplicity, we consider only resistive termination impedances. We may then represent the circuit as shown in Figure 3-10. To avoid confusion, we choose the direction of I_2 and the polarity of V_N such that their real parts will be positive when V_S is positive. It is important to note that the polarity of inductively coupled noise is opposite to that of capacitively coupled noise, as will become apparent from the solution.

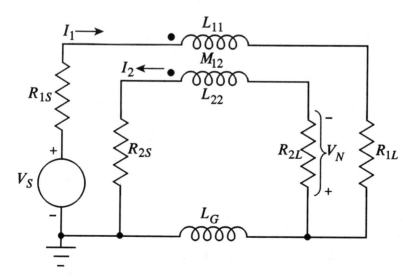

Figure 3-10   Inductances in two signal paths as actual elements

The loop equations for this circuit are

$$(R_{IS} + j\omega L_{11} + R_{1L})I_1 - j\omega M_{12}I_2 + j\omega L_G(I_1 - I_2) = V_S$$

$$[R_{1S} + R_{1L} + j\omega(L_{11} + L_G)]I_1 - j\omega(M_{12} + L_G)I_2 = V_S \qquad \textbf{(3-42)}$$

$$[R_{2S} + j\omega L_{22} + R_{2L}]I_2 - j\omega M_{12}I_1 + j\omega L_G(I_2 - I_1) = 0$$

$$-j\omega(M_{12} + L_G)I_1 + [j\omega(L_{22} + L_G) + R_{2S} + R_{2L}]I_2 = 0 \qquad \textbf{(3-43)}$$

Note that wherever L_G appears, it is always added to either L_{11}, L_{22}, or M_{12}. Since M_{12} is part of L_{11} and L_{22}, L_G is always added to M_{12} and their individual effects are indistinguishable. Any flux linking both loops must have the same effect whether it is due to M_{12} or L_G. This will be an important result later when we discuss grounding techniques.

To simplify the notation, let $L_1 = L_{11} + L_G$, $L_2 = L_{22} + L_G$, $L_{12} = M_{12} + L_G$, $R_1 = R_{1S} + R_{1L}$, and $R_2 = R_{2S} + R_{2L}$. Physically, L_1 results from all flux created by the current I_1 that links loop 1, and similarly for L_2. The inductance L_{12} results from all flux that links both loops. By using these designations and the equivalent circuit for a transformer, we may redraw Figure 3-10 as shown in Figure 3-11.

Solving Equations (3-42) and (3-43) for I_2 yields

$$I_2 = \frac{j\omega L_{12}V_S}{(R_1 + j\omega L_1)(R_2 + j\omega L_2) - (j\omega L_{12})^2} \qquad \textbf{(3-44)}$$

The noise voltage V_N is then equal to $I_2 R_{2L}$, or

$$V_N = \frac{j\omega L_{12}R_{2L}V_S}{(R_1 + j\omega L_1)(R_2 + j\omega L_2) - (j\omega L_{12})^2}$$

$$= \frac{j\omega L_{12}R_{2L}V_S}{(j\omega)^2(L_1L_2 - L_{12}^2) + j\omega(L_2R_1 + L_1R_2) + R_1R_2} \qquad \textbf{(3-45)}$$

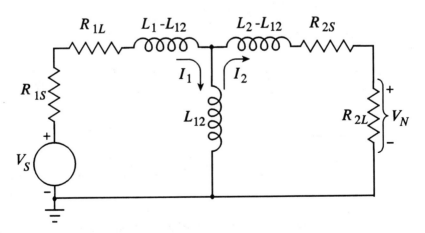

Figure 3-11   Inductances in two signal paths as a wye network

Like the earlier expression for capacitive coupling, this expression contains two real poles and a zero, which may be seen by factoring the denominator:

$$V_N = \frac{L_{12}R_{2L}V_S}{L_1L_2 - L_{12}^2} \cdot \frac{j\omega}{(j\omega + s_c)(j\omega + s_d)} \tag{3-46}$$

where

$$s_c, s_d = \frac{L_2R_1 + L_1R_2 \pm \sqrt{(L_2R_1 - L_1R_2)^2 + 4L_{12}^2R_1R_2}}{2(L_1L_2 - L_{12}^2)}$$

$$= \frac{L_1}{2R_1} + \frac{L_2}{2R_2} \mp \frac{1}{\sqrt{\left(\dfrac{L_1}{2R_1} - \dfrac{L_2}{2R_2}\right)^2 + \dfrac{L_{12}^2}{R_1R_2}}} \tag{3-47}$$

We can see that the behavior of inductive coupling is similar to capacitive coupling, but with the frequency inverted. At low frequencies, the $j\omega$ terms in the denominator become insignificant, and Equation (3-46) reduces to

$$V_N = \frac{L_{12}R_{2L}V_S}{L_1L_2 - L_{12}^2} \cdot \frac{j\omega}{s_c s_d} = j\omega \frac{L_{12}R_{2L}V_S}{R_1R_2} \tag{3-48}$$

So at low frequencies, the resistances dominate the inductive reactances, and V_S/R_1 behaves as a constant current source. The inductances act as a voltage divider, and I_2 and V_N are proportional to frequency.

At high frequencies, the s_c and s_d terms in the denominator of Equation (3-46) become insignificant, and the equation reduces to

$$V_N = \frac{L_{12}R_{2L}V_S}{L_1L_2 - L_{12}^2} \cdot \frac{j\omega}{(j\omega)^2} = \frac{L_{12}R_{2L}V_S}{j\omega(L_1L_2 - L_{12}^2)} \tag{3-49}$$

At these frequencies, all resistances become insignificant, the inductances behave as a current divider, and I_1, I_2, and V_N are inversely proportional to frequency.

As with capacitive coupling, the mid-range behavior of Equation (3-46) becomes important if s_c and s_d are far apart. This occurs if L_1/R_1 is considerably different from L_2/R_2, or if L_{12} is nearly equal to L_1 and L_2. In the former case, if R_2 appears as an open circuit while R_1 appears as a short, then $L_1 - L_{12}$ forms a voltage divider with L_{12}. Conversely, if R_1 is very high and R_2 very low, then $L_2 - L_{12}$ acts as a current divider with L_{12}. For the case where L_{12} is nearly equal to L_1 and L_2, the inductances behave as a perfectly coupled transformer. Then L_{12} has a high reactance in the middle frequency range, and R_1 and R_2 form a voltage divider. In all three cases, V_N is nearly constant over the mid-range frequencies. At these frequencies, ω dominates over s_c but is insignificant compared to s_d. Then Equation (3-46) reduces to

$$V_N = \frac{L_{12}R_{2L}V_S}{(L_1L_2 - L_{12}^2)s_d} \tag{3-50}$$

where s_d is as in Equation (3-47). This value of V_N is the maximum inductively coupled noise voltage over all frequencies.

For example, assume $R_1 = R_2 = 80\ \Omega$, $L_{12} = 2.3\ \mu H$, and $L_1 = L_2 = 2.76\ \mu H$, which are typical values. Then from Equation (3-6), $s_c = 15.8 \cdot 10^6$ rad/s or 2.51 MHz, and $s_d = 173.7 \cdot 10^6$ rad/s or 27.6 MHz. If $R_{2L} \gg R_{2S}$ so that $R_2 \approx R_{2L}$, Equation (3-45) or (3-46) states that at $f = 10$ MHz (where $\omega = 2\pi f$), $|V_N| = 0.415\ |V_S|$. Since $s_c < 10$ MHz $< s_d$, Equation (3-50) should give a reasonable approximation for $|V_N|$; it does, since it says that $|V_N| \approx 0.455\ |V_S|$. The equivalence of this $|V_N|$ value to the mid-range value obtained for capacitive coupling is not coincidental. It follows from the fact that the inductances result from the same physical geometry as the capacitances, as we will study shortly.

The plot of $|V_N/V_S|$ versus frequency, for the above R and L values, appears in Figure 3-12. The straight-line approximations and the corner frequencies of the $L_{12} = 2.3\ \mu H$ curve are superimposed on the graph. Curves are also shown for other values of L_{12}, with L_1 and L_2 still equal to 2.76 μH.

As with the earlier equation for capacitive coupling, we can plot a more general graph by defining four dimensionless parameters. For inductive coupling we define them as follows:

$$P_f = \omega \sqrt{\frac{L_1 L_2}{R_1 R_2}} \qquad\qquad P_{12} = \frac{L_{12}}{\sqrt{L_1 L_2}}$$

$$P_D = \frac{L_1 R_2 + L_2 R_1}{2\sqrt{L_1 L_2 R_1 R_2}} \qquad\qquad P_V = \frac{V_N \sqrt{R_1 R_2}}{V_S R_{2L}} \tag{3-51}$$

Equation (3-45), in terms of these parameters, then becomes

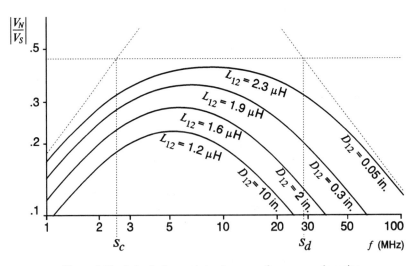

Figure 3-12   Inductively coupled noise versus frequency and spacing

$$P_V = \frac{jP_fP_{12}}{(jP_f)^2(1 - P_{12}^2) + j2P_fP_D + 1} \tag{3-52}$$

For identical loops, that is, for $L_1 = L_2$ and $R_1 = R_2$ as in the above example, $P_D = 1$. A plot of P_V versus P_f, for $P_D = 1$ and for P_{12} calculated from the values of L_{12} in the above example, appears in Figure 3-13. The corner frequency points, also appearing in the figure for $P_{12} = 0.833$, are

$$P_{fc}, P_{fd} = \frac{P_D \pm \sqrt{P_D^2 - 1 + P_{12}^2}}{1 - P_{12}^2} \tag{3-53}$$

or, since $P_D = 1$,

$$P_{fc}, P_{fd} = \frac{1 \pm \sqrt{1 - 1 + P_{12}^2}}{1 - P_{12}^2} = \frac{1 \pm P_{12}}{1 - P_{12}^2} = \frac{1}{1 \pm P_{12}} \tag{3-54}$$

We can use this graph for any geometry for which $P_D = 1$. For $R_{1P} = R_{2P} = 80\ \Omega$, and for $L_1 = L_2 = 2.76\ \mu\text{H}$, from the example,

$$P_f = \omega \sqrt{\frac{2.76\ \mu\text{H} \cdot 2.76\ \mu\text{H}}{80\ \Omega \cdot 80\ \Omega}} = \omega \cdot 34.5\ \text{ns} = 0.217 \cdot f \tag{3-55}$$

with f in MHz. This verifies the correspondence between the scales of Figures 3-12 and 3-13.

For other values of R and L, we first calculate P_D. If $P_D = 1$, we can use Figure 3-13; otherwise, we must plot a different graph. If $P_D = 1$, we calculate P_{12} to find

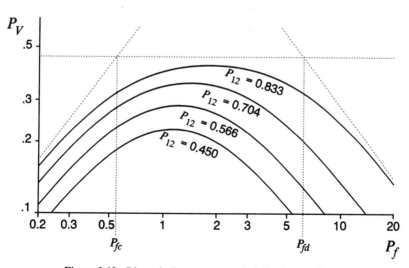

Figure 3-13 Dimensionless parameters for inductive coupling

which curve in Figure 3-13 is nearest to it. Then we calculate the coefficient of f in P_f, to scale the figure. For example, suppose R_1 and R_2 are 400 Ω instead of 80 Ω but the inductances remain the same. Then, for the same frequency, P_f decreases by a factor of five, as in the capacitive example of Section 3-1. Again, the effect is to increase the frequency scale of Figure 3-13 by this factor. Thus, with inductive coupling, as the termination resistances increase, the frequency range of maximum concern ($s_c < 2\pi f < s_d$) moves upward. When this frequency range becomes greater than 1000 MHz, the analysis is no longer applicable. Inductive coupling is therefore of greatest concern for *low* termination impedances.

In the above inductive example, we also assumed that $R_{2L} \gg R_{2S}$. If this is not true (but still $R_1 = R_2$), then the factor of $\sqrt{R_1 R_2 / R_{2L}}$ is no longer approximately equal to 1. This changes the voltage scale in Figure 3-13 by the factor of $\sqrt{R_1 R_2 / R_{2L}}$, according to the definition of P_V for inductive coupling.

The inductive coupling can be modeled as shown in Figure 3-14, where the voltage source V_2 is the voltage induced into loop 2 by the current I_1. By solving Equations (3-42) and (3-43) for I_1, we show the magnitude of V_2 to be

$$V_2 = j\omega L_{12} I_1 = j\omega L_{12} \frac{R_2 + j\omega L_2}{(R_1 + j\omega L_1)(R_2 + j\omega L_2) - (j\omega L_{12})^2} V_S \qquad (3\text{-}56)$$

Inductance can be reduced by separating the leads or by twisting the source leads if the return current passes through another lead instead of through the ground. To quantify this effect, we must calculate the inductance.

· We can find the inductance in a loop consisting of two parallel wires in a manner similar to that used for finding the capacitance between them. Again, the values will be only approximate, but they will suggest how we can reduce the inductive coupling. Consider again the two parallel wires shown in Figure 3-9, and

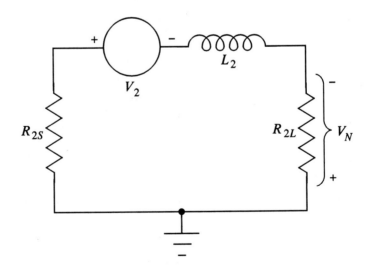

Figure 3-14   Inductive noise model

assume that they are connected at each end. Assume that a current I_{ab} is flowing through lead a in the $+z$ direction and returning through lead b. The magnetic vector potential $\mathbf{A}$ is a useful concept[2,3] that, for a two-dimensional geometry, behaves much like the electric scalar potential V. Its curl is equal to $\mathbf{B}$, and it satisfies an equation similar to Poisson's equation:

$$\nabla^2 \mathbf{A} = -\mu \mathbf{J} \qquad (3\text{-}57)$$

where $\nabla^2 \mathbf{A} = \mathbf{i}\nabla^2 A_x + \mathbf{j}\nabla^2 A_y + \mathbf{k}\nabla^2 A_z$, $\mathbf{i}$, $\mathbf{j}$, and $\mathbf{k}$ are unit vectors in the x, y, and z directions, μ is the permeability, and $\mathbf{J}$ is the current density. Most insulation materials are non-magnetic. For such materials, the permeability is the same as that for free space, designated μ_0:

$$\mu = \mu_0 = 4\pi \cdot 10^{-7} \text{ H/m} \approx 1.257 \ \mu\text{H/m} \qquad (3\text{-}58)$$

Since all current is flowing in the $\pm z$ direction, both J_x and J_y are zero, and so are A_x and A_y. Thus $\mathbf{A}$ is always normal to the page in Figure 3-9.

    If the conductors are perfect, then skin effect causes the current to be concentrated at the conductor surfaces. Also, the magnetic flux density $\mathbf{B}$ must be tangential to the conductor surface. Otherwise, the current would be deflected sideways along the surface and would redistribute itself such as to make $\mathbf{B}$ tangential. Since $\mathbf{B} = \nabla \times \mathbf{A}$,

$$\mathbf{B} \cdot (\nabla A_z) = (\nabla \times \mathbf{A}) \cdot (\nabla A_z)$$

$$= \left(\mathbf{i}\frac{\partial A_z}{\partial y} - \mathbf{j}\frac{\partial A_z}{\partial x} \right) \cdot \left(\mathbf{i}\frac{\partial A_z}{\partial x} + \mathbf{j}\frac{\partial A_z}{\partial y} \right)$$

$$= \frac{\partial A_z}{\partial y} \frac{\partial A_z}{\partial x} - \frac{\partial A_z}{\partial x} \frac{\partial A_z}{\partial y} = 0 \qquad (3\text{-}59)$$

and therefore $\mathbf{B}$ is normal to the gradient of A_z, or tangential to its equipotential surfaces. Thus, each conductor surface coincides with an equipotential surface of $\mathbf{A}$. The current distributes itself in the conductor surface such that the surface attains a constant vector potential $\mathbf{A}$.

    Equations (3-57) and (3-18) are of the same form, and their boundary conditions are identical. Thus, their solutions must be of the same form. The solution of Equation (3-57) is

$$\mathbf{A}(x, y) = A_0 \ln \left[\frac{(x + \sqrt{s^2 - r^2})^2 + y^2}{(x + \sqrt{s^2 - r^2})^2 + y^2} \right] \mathbf{k} \qquad (3\text{-}60)$$

where A_0 is a scale factor similar to V_0 in Equation (3-17).

---

    2. William H. Hayt, *Engineering Electromagnetics*, 5th ed. (New York, N.Y.: McGraw-Hill Publishing Company, 1989), pp. 249–57.

    3. John D. Kraus, *Electromagnetics*, 4th ed. (New York, N.Y.: McGraw-Hill Publishing Company, 1992), pp. 258–63.

The constant magnetic vector potential $\mathbf{A}_a$ at the surface of conductor a is

$$\mathbf{A}_a = 2A_0 \ln\left[\sqrt{\left(\frac{s}{r}\right)^2 - 1} + \frac{s}{r}\right]\mathbf{k} = 2A_0 \cosh^{-1}\left(\frac{s}{r}\right)\mathbf{k} \qquad (3\text{-}61)$$

and at the surface of conductor b it is the negative of this value. The two conductors, connected at each end, form a long closed loop. The total magnetic flux Φ, enclosed by the loop is

$$\Phi = \int_S \mathbf{B} \cdot d\mathbf{s} = \int_S (\nabla \times \mathbf{A}) \cdot d\mathbf{s} \qquad (3\text{-}62)$$

where S is the surface of the conductors forming the loop. The loop encloses the strip of the xz plane between the conductors. By Stokes's theorem, the surface integral of $\nabla \times \mathbf{A}$ over that strip is equal to the line integral of $\mathbf{A}$ around the loop. Although $\mathbf{A}$ varies over the strip, it is constant and equal to $\mathbf{A}_a$ everywhere on the surface of conductor a, and equal to $-\mathbf{A}_a$ at the surface of conductor b. Therefore, if we neglect the ends of the loop,

$$\Phi = \int_S (\nabla \times \mathbf{A}) \cdot d\mathbf{s} = \oint \mathbf{A} \cdot d\mathbf{l} = 2|\mathbf{A}_1|l = 4A_0 l \cosh^{-1}\left(\frac{s}{r}\right) \qquad (3\text{-}63)$$

where l is the length of each conductor.

To find the inductance of this loop, we must know the current I_{ab} in the loop. By Ampere's law, it is equal to the integral of the magnetic field intensity, $\mathbf{H}$, around any closed path linking the loop. We choose a path consisting of a section of the y axis, with its ends connected by a semicircle of infinite radius. The line integral of $\mathbf{H}$ over the semicircle is zero, since $|\mathbf{H}|$ decreases inversely with the square of the radius. Thus, the only non-zero part of the integral is on the y axis and is

$$I_{ab} = \oint \mathbf{H} \cdot d\mathbf{l} = -\int_{-\infty}^{\infty} H_y\Big|_{x=0} dy = \frac{1}{\mu}\int_{-\infty}^{\infty} \frac{\partial A_z}{\partial x}\Big|_{x=0} dy \qquad (3\text{-}64)$$

Due to the similarity between the respective functions for $\mathbf{A}$ and V, $\partial A_z/\partial x$ resembles the expression in Equation (3-20). Substituting the analogous expression into Equation (3-64) yields

$$I_{ab} = \frac{1}{\mu}\int_{-\infty}^{\infty} A_0 \left[\frac{2(0 + \sqrt{s^2 - r^2})}{(0 + \sqrt{s^2 - r^2})^2 + y^2} - \frac{2(0 - \sqrt{s^2 - r^2})}{(0 - \sqrt{s^2 - r^2})^2 + y^2}\right] dy$$

$$= \frac{4A^0}{\mu} \tan^{-1}\frac{y}{\sqrt{s^2 - r^2}}\Bigg]_{-\infty}^{\infty} = \frac{4\pi A_0}{\mu} \qquad (3\text{-}65)$$

The inductance L_{ab} of this loop is equal to the flux linkage divided by the current, and since the loop consists of only one turn, this is the flux Φ, divided by the current I_{ab}:

$$L_{ab} = \frac{\Phi}{I_{ab}} = \frac{4A_0 l \cosh^{-1}\left(\frac{s}{r}\right)}{4\pi A_0/\mu} = \frac{\mu l}{\pi} \cosh^{-1}\left(\frac{s}{r}\right) \qquad (3\text{-}66)$$

Here also, if d is the wire diameter and D_{ab} is the spacing, then since $d = 2r$ and $D_{ab} = 2s$, we may express the formula more conveniently as follows:

$$L_{ab} = \frac{\mu l}{\pi} \cosh^{-1} \left(\frac{D_{ab}}{d} \right) \tag{3-67}$$

Also, if the wire spacing is much larger than the diameter,

$$L_{ab} \approx \frac{\mu l}{\pi} \ln \left(\frac{2D_{ab}}{d} \right) \text{ for } D_{ab} \gg d \tag{3-68}$$

We again consider two #30 AWG wires ($d = 0.01$ in.) separated by 5 in., in free space, for which $D_{ab}/d = 500$. From Equation (3-68), for a length of 1 m, $L_{ab} = 2.76$ μH, the value assumed earlier in Figure 3-12. Therefore, the inductances assumed for Figure 3-12 occur if the distances D_{1G} and D_{2G} are 5 in., with 1-m lengths of #30 AWG wire.

If the ground return in Figure 3-10 is another lead of the same diameter d, then to find inductance L_1 we assign lead 1 to lead a and the ground return to lead b. We set D_{ab} equal to D_{12}.

$$L_1 = \frac{\mu l}{\pi} \cosh^{-1} \left(\frac{D_{1G}}{d} \right) \approx \frac{\mu l}{\pi} \ln \left(\frac{2D_{1G}}{d} \right) \text{ for } D_{1G} \gg d \tag{3-69}$$

We find inductance L_2 using the same procedure.

$$L_2 = \frac{\mu l}{\pi} \cosh^{-1} \left(\frac{D_{2G}}{d} \right) \approx \frac{\mu l}{\pi} \ln \left(\frac{2D_{2G}}{d} \right) \text{ for } D_{2G} \gg d \tag{3-70}$$

If the return path is a ground plane instead of another lead, the fields behave as though there were an image of lead 1 below the plane. If h_1 is the height of lead 1 above the plane, then by reasoning similar to that used for capacitance, the formula for inductance L_1 is

$$L_1 = \frac{\mu l}{2\pi} \cosh^{-1} \left(\frac{2h_1}{d} \right) \approx \frac{\mu l}{2\pi} \ln \left(\frac{4h_1}{d} \right) \text{ for } h_1 \gg d \tag{3-71}$$

A similar expression applies to L_2.

To find L_{12}, we must calculate the flux linking loop 2 that is generated by the current I_1. This requires a solution to Equation (3-57) that satisfies the boundary conditions consisting of all three conductors. The solution given by Equation (3-60) is valid only for two conductors. If we introduce a third finite-sized conductor, it changes the magnetic boundary conditions in much the same way as the electric boundary conditions discussed earlier. Then Equation (3-60) is no longer valid. A more complex solution is not warranted here either. However, if the conductor diameters are much less than their spacing, then any new conductors parallel to the z axis do not affect the magnetic vector potential. The expression of Equation (3-60) then still applies. To simplify the solution, it will again be assumed that $d \ll D_{1G}$ and $d \ll D_{2G}$, or equivalently, $r \ll s$ in both cases.

Now we assign lead a in Figure 3-9 to lead 1, and lead b to the ground return. Substituting Equation (3-65) into Equation (3-60) with $r \ll s$ shows that at any point (x, y), the magnetic vector potential due to the current in loop 1 is

$$\mathbf{A}(x, y) = \frac{\mu I_1}{4\pi} \ln \left[\frac{(x+s)^2 + y^2}{(x-s)^2 + y^2} \right] \mathbf{k} \tag{3-72}$$

where s now represents $D_{1G}/2$. If lead 2, parallel to the z axis, is added at (x_2, y_2), the magnetic vector potential $\mathbf{A}_2$ along that lead is

$$\mathbf{A}_2 = \frac{\mu I_1}{4\pi} \ln \left[\frac{(x_2+s)^2 + y_2^2}{(x_2-s)^2 + y_2^2} \right] \mathbf{k}$$

$$= \frac{\mu I_1}{4\pi} \ln \left(\frac{D_{2G}^2}{D_{12}^2} \right) \mathbf{k} = \frac{\mu I_1}{2\pi} \ln \left(\frac{D_{2G}}{D_{12}} \right) \mathbf{k} \tag{3-73}$$

We find the magnetic vector potential $\mathbf{A}_G$ at the surface of the ground return conductor by substituting Equation (3-65) into Equation (3-61) and recalling that $\mathbf{A}_G = -\mathbf{A}_1$:

$$\mathbf{A}_G = -\frac{\mu I_1}{2\pi} \cosh^{-1} \left(\frac{D_{1G}}{d} \right) \mathbf{k} \approx -\frac{\mu I_1}{2\pi} \ln \left(\frac{2D_{1G}}{d} \right) \mathbf{k} \text{ for } D_{1G} \gg d \tag{3-74}$$

The flux Φ_{12} enclosed by loop 2 due to the current in loop 1 is

$$\Phi_{12} = \int_{S_2} \mathbf{B} \cdot ds = \oint \mathbf{A} \cdot dl = |\mathbf{A}_2| l + |\mathbf{A}_G| l$$

$$= \frac{\mu I_1 l}{2\pi} \left[\ln \left(\frac{D_{2G}}{D_{12}} \right) + \ln \left(\frac{2D_{1G}}{d} \right) \right] = \frac{\mu I_1 l}{2\pi} \ln \left(\frac{2D_{1G}D_{2G}}{dD_{12}} \right) \tag{3-75}$$

The mutual inductance L_{12} is

$$L_{12} = \frac{\Phi_{12}}{I_1} = \frac{\mu l}{2\pi} \ln \left(\frac{2D_{1G}D_{2G}}{dD_{12}} \right) \tag{3-76}$$

With this expression, we can now verify the correspondence between L_{12} and D_{12} in Figure 3-12, where $D_{1G} = D_{2G} = 5$ in. and $d = 0.01$ in.

By substituting Equations (3-69), (3-70), and (3-76) into Equation (3-45), one obtains the expression for V_N in terms of D_{1G}, D_{12}, and D_{2G}:

$$V_N = \frac{\dfrac{2\pi R_{2L}}{j\omega \mu l} \cdot \ln \dfrac{2D_{1G}D_{2G}}{dD_{12}}}{4 \left(\dfrac{\pi R_1}{j\omega \mu l} + \ln \dfrac{2D_{1G}}{d} \right) \left(\dfrac{\pi R_2}{j\omega \mu l} + \ln \dfrac{2D_{2G}}{d} \right) - \left(\ln \dfrac{2D_{1G}D_{2G}}{dD_{12}} \right)^2} V_S \tag{3-77}$$

Consider an example similar to that used for the capacitive case, of two #30 AWG wires, designated 1 and 2, in free space, each being 1 m long. The distance between the wires is 2 in., and the common #30 AWG ground return is separated from each lead by 5 in. This is the geometry assumed for Figure 3-12, so $L_1 = L_2 = 2.76$ μH and $L_{12} = 1.6$ μH. Lead 1 is driven by a 5-V, 50-MHz sine wave signal. Now, however, let the source and load impedances on both leads be equal to 100 Ω each. The resistances R_1 and R_2 of their series combinations are each then 200 Ω.

Since R_1 and R_2 are 80 Ω for the graph in Figure 3-12, we cannot directly use Figure 3-12, but, since $P_D = 1$, we can use Figure 3-13 by calculating P_{12} and the coefficients of P_f and P_V. From the inductances mentioned above, we find that $P_{12} = 0.566$. Since $f = 50$ MHz, $P_f = 2\pi \cdot (50 \cdot 10^6) \cdot (2.76 \cdot 10^{-6})/200 = 4.34$. Thus we can read P_V from the $P_{12} = 0.566$ curve in Figure 3-13, at $P_f = 4.34$. Its value is 0.17, and from this we calculate $V_N = 0.17 \cdot (5 \text{ volts}) \cdot 100 \ \Omega / 200 \ \Omega = 0.425$ volts. The P_f value of 4.34 is above the peak frequency, so if the frequency decreases, V_N increases until its peak is reached, after which V_N decreases as shown.

If the return path is a ground plane then we must assume images of leads 1 and 2 below the ground plane. We then use Equation (3-71) to calculate L_1 and L_2. The resulting graph will still resemble Figure 3-12.

Comparing Equations (3-41) and (3-77) shows that inductive and capacitive coupling behave very similarly as the geometry of a two-dimensional pattern varies. Note, however, that capacitively coupled noise is proportional to R_{2P}, the parallel impedances of the source and load on the "sense" lead. Inductively coupled noise, on the other hand, is proportional to the ratio of R_{2L}, the load impedance, to R_2, the total series impedance. Therefore, to learn whether noise is inductively or capacitively coupled, we may decrease R_{2S}, the source impedance on the sense lead. We may do this by connecting a resistor in parallel with R_{2S}, or if this disturbs circuit operation, we may replace the driver circuit with one of lower impedance. Capacitively coupled noise would then decrease, whereas inductively coupled noise would increase.

Although we calculated the mutual inductance for the parallel case, it exists for every geometry. Unfortunately, for shapes that are not constant along one axis, calculation is much more complex. The magnetic vector potential $\mathbf{A}$ is no longer everywhere in the same direction and therefore cannot be treated the same as the electric scalar potential. Therefore, there is no simple correspondence between capacitance and inductance. However, it is still true that

$$V_N = \frac{d}{dt} \int_S \mathbf{B} \cdot d\mathbf{S} = j\omega BS \cos\theta \tag{3-78}$$

for constant sinusoidal $\mathbf{B}$ and a flat surface $\mathbf{S}$. Obviously we can reduce V_N by reducing the area of $\mathbf{S}$ or by increasing θ, the angle between $\mathbf{B}$ and $\mathbf{S}$, toward 90°.

3-3 PROBLEMS

1. The circuit of Figure 3-3 contains two nodes, connected by C_{12}, plus the ground reference node. Write the node equations for this circuit, and solve them to show that

$$V_N = \frac{j\omega C_{12} / R_{1S}}{[j\omega(C_{12} + C_{1G}) + 1/R_{1P}][j\omega(C_{12} + C_{2G}) + 1/R_{2P}] - (j\omega C_{12})^2} V_S \quad (3\text{-}79)$$

where R_{1P} represents the parallel combination of R_{1S} and R_{1L}, and similarly for R_{2P}.

2. The three impedances of a wye circuit are designated Z_1, Z_2, and Z_3, and are connected to nodes 1, 2, and 3, respectively, with their other ends joined. The three impedances of the equivalent delta circuit are designated Z_{12}, Z_{13}, and Z_{23}, each connected between the two nodes denoted by its subscripts. The delta impedance Z_{12} is related to the wye impedances by the formula

$$Z_{12} = \frac{Z_1 Z_2 + Z_2 Z_3 + Z_3 Z_1}{Z_3} \quad (3\text{-}80)$$

(a) Use symmetry to derive the formulae for Z_{13} and Z_{23} in terms of Z_1, Z_2, and Z_3.

(b) Let Z_{12}, Z_{13}, and Z_{23} represent the impedances of C_{12}, C_{1G}, and C_{2G}, respectively. Substitute the wye equivalent formulae [Equation (3-80)] into Equation (3-79) and show that it yields the same noise voltage as the wye circuit of Equation (3-4).

3. Let C_{1GT} be the total measured or calculated capacitance between leads 1 and G. As explained in the text, it is equal to C_{1G} in parallel with the series combination of C_{12} and C_{2G}. Similarly, let C_{12T} and C_{2GT} be the total capacitance between leads 1 and 2 and between leads 2 and G, respectively. By using the formulae for capacitances in series and parallel, show that

$$C_{1G} = \frac{2\left(\dfrac{1}{C_{12T}} + \dfrac{1}{C_{2GT}} - \dfrac{1}{C_{1GT}}\right)}{\dfrac{4}{C_{12T} C_{2GT}} - \left(\dfrac{1}{C_{12T}} + \dfrac{1}{C_{2GT}} - \dfrac{1}{C_{1GT}}\right)^2} \quad (3\text{-}81)$$

Derive similar expressions for C_{12} and C_{2G}, and check them by symmetry.

4. Use Equations (3-35) and (3-81) to derive an expression for C_{1G} in terms of the diameter and spacing of leads 1, 2, and G. Repeat for C_{12} and C_{2G} using the similar expressions derived in Problem 3. Substitute these expressions into Equation (3-79) to show that the resulting expression for V_N, in terms of the geometric dimensions of the leads, is equivalent to Equation (3-41).

5. Let the wires in Figure 3-9 be of unequal radii r_a and r_b, respectively, with their centers at $x = +s$ and $x = -s$. Show that the potential $V(x, y)$ in the space around the wires varies according to the following function:

$$V(x, y) = V_0 \ln \left[\frac{(x + k + p)^2 + y^2}{(x + k - p)^2 + y^2}\right] + V_k \quad (3\text{-}82)$$

where

$$k = \frac{r_a^2 - r_b^2}{4s}, \quad p = \sqrt{s^2 - \frac{r_a^2 + r_b^2}{2} + k^2}$$

and where V_0 and V_k are constants that depend on the potentials on the wires.

6. Derive the formula for the potential on each wire in Problem 5 in terms of V_0, V_k, r_a, r_b, and s. Assume that the potentials on the wires are $+1/2$ volt and $-1/2$ volt, respectively, and solve for the values of V_0 and V_k in terms of r_a, r_b, and s.

7. Derive the formula for the capacitance between the two wires in Problem 5. Show that if $r_a = r_b$, it reduces to Equation (3-33).

8. Two wires are located a height h above a conducting plane and are separated by a distance of $2s$. Their diameter d is much less than h and s. A potential of $+1$ V is applied to one of the leads, and the other lead is left unconnected. The potential on the conducting plane is 0 V. Find an expression for the potential on the floating lead, in terms of h, s, and d.

9. The equivalent circuit of Problem 8 is Figure 3-4, with $R_{2S} = R_{2L} = \infty$. Since no current can flow through C_2, there is no potential difference across it, and the potential across C_G is the potential on the floating lead. Also, the charge on C_G must be the same as the charge on C_1 and C_G in series. Use this information and the result of Problem 8 to derive a formula for the capacitance C_G, in terms of h, s, and d.

10. Derive the formula for the inductance between the two wires of Problem 5 by using an expression for $A(x, y)$ like the expression for V. Show that if $r_a = r_b$, it reduces to Equation (3-66).

4

SHIELDING OF CABLES TO REDUCE CAPACITIVE AND INDUCTIVE COUPLING

In situations where we cannot sufficiently reduce coupled noise by separating the conductors, we can usually reduce it further by shielding them. We shall now investigate the effects of shielded cable on capacitively and inductively coupled noise. The shield consists of a metallic braid or a metallized sleeve, which we will approximate by a solid conductive tube surrounding the wire. When the shield is properly connected, electric charges flow onto the shield and generate an electric field that nullifies the capacitively coupled noise. Currents also flow in the shield and generate a magnetic field that nullifies the inductively coupled noise.

We may shield either the source lead or the victim lead, or in extreme cases, both. With only one lead shielded, the same attenuation will result no matter which lead it is. This is true because both transfer functions must be equal due to reciprocity. If there is one noisy source lead affecting several sensitive victim leads, then it is easier to shield only the noisy lead. Conversely, if a victim lead is particularly sensitive and several source leads are affecting it, then it is usually easier to shield the sensitive lead.

4-1 REDUCTION OF CAPACITIVE (ELECTRIC) COUPLING

We showed earlier that the equipotential surfaces between two parallel wires are cylinders enclosing either of the wires. These surfaces appear in Figure 4-1, which also shows a shield around lead 1. One such equipotential surface has the same diameter as the shield. If the shield radius is much smaller than the lead spacing, then even without the shield, the lead axis is approximately at the axis of that

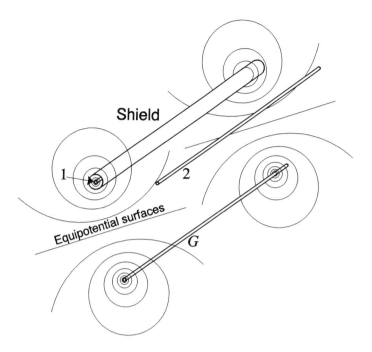

Figure 4-1 Electric equipotential surfaces around cable shield

equipotential surface. When we add the shield, its surface coincides with a surface whose potential is already constant. Thus, it does not change the potential or distort the electric field. If the shield does not connect to anything, it cannot have any net charge and therefore cannot add any new fields. Thus, since the shield does not affect the field in any way, it has no effect on the capacitive coupling, and it is therefore useless in that respect. We will use it only to develop an equivalent circuit that will later be usable with a grounded shield.

The shield behaves like a thicker lead, and any potential on it couples to lead 2 and lead G through the electric field outside the shield. We may therefore use the wye equivalent circuit for the capacitances between the shield, lead 2, and lead G. Let the wye circuit capacitances be C_s, C_2, and C_G, each connected to the node designated by its subscript. Let C_{1s} be the capacitance between lead 1 and the shield. Since the shield is floating and has no net charge, any displacement current (dQ/dt) flowing through C_{1s} from lead 1 onto the shield must continue into the space outside the shield. We consider this current to flow into the equivalent wye circuit through C_s. Capacitance C_{1s} is thus in series with C_s. The equivalent circuit is shown in Figure 4-2.

To find C_s and C_G, we temporarily assume lead 2 to be floating ($R_{2S} = R_{2L} = \infty$), as we did in Chapter 3. Capacitor C_s is then in series with C_G. If we let $S_s = 1/C_s$ and $S_G = 1/C_G$, then the total elastance between the shield and lead G is

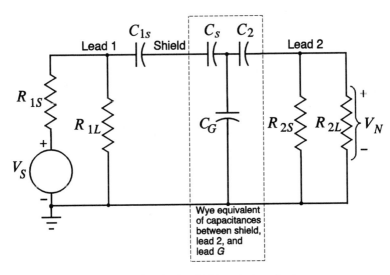

Figure 4-2 Floating shield, equivalent circuit

$S_s + S_G$. This may be calculated from the physical dimensions, like $S_1 + S_G$ in Figure 3-4, assuming that lead 2 is very thin, as before. The total elastance between lead 1 and lead G is $S_{1s} + S_s + S_G$, where $S_{1s} = 1/C_{1s}$, which we will calculate shortly. When we introduce the floating shield, the series combination of C_{1s} and C_s thus replaces C_1 of Figure 3-4.

If we now connect the shield to lead G, a charge opposite to that on the center lead will flow onto the shield, as shown in Figure 4-3. Both charges normally vary with time. If the connection is perfect and contains no inductance, the charges will always be equal and opposite. The net electric field outside the shield will then be zero, so no noise will be capacitively coupled onto lead 2. Capacitive coupling may

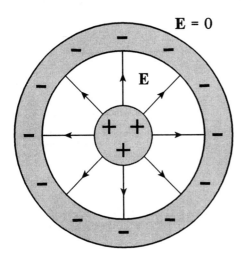

Figure 4-3 Charges and electric fields in shielded cable

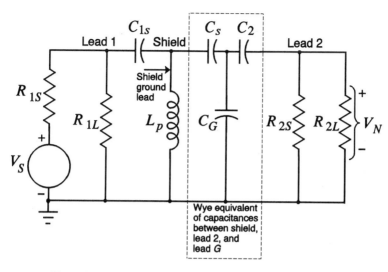

Figure 4-4   Capacitive coupling through shield, equivalent circuit

therefore be effectively reduced in this manner. The equivalent circuit appears in Figure 4-4, but with L_p replaced by a short circuit ($L_p = 0$) for this ideal case.

In a real circuit, the connection between the shield and lead G is often a short lead commonly known as a *pigtail* lead. Unfortunately, this lead contains inductance, designated L_p in Figure 4-4. Its effects are easiest to analyze using circuit theory. Using the methods of Chapter 3 shows that, if L_p is significant, the shield loses its effect at high frequencies. The pigtail lead should be kept as short and thick as possible in order to minimize this effect.

Another concern is that part of the center lead must usually extend beyond the shield to connect it to its termination. The exposed lead behaves like any other unshielded lead; therefore, the exposed length should be kept to a minimum. We may calculate the capacitive coupling to the exposed part by using the methods of Chapter 3, letting l represent the length of only the unshielded portion.

To compute V_N/V_S as a function of the physical dimensions, we must know the capacitances. All were calculated in Chapter 3 except C_{1s}, which we shall now calculate. Assume a potential of $+V_{1s}/2$ on lead 1 and $-V_{1s}/2$ on the shield, so that the potential difference between them is V_{1s}. Let their respective radii be r_1 and r_s. Let (ρ, Φ, z) be the cylindrical coordinates of an arbitrary point between the center conductor and the shield. To calculate C_{1s}, we must know the potential $V(\rho)$ at every such point. By solving Laplace's equation and testing the boundary conditions, we may show that this potential is

$$V(\rho) = V_{1s}\frac{\ln\left(\rho/\sqrt{r_1 r_s}\right)}{\ln\left(r_1/r_s\right)} \tag{4-1}$$

Then by calculating the electric flux over surface of lead 1 and using Gauss's law, it follows that the capacitance C_{1s} is

$$C_{1s} = \frac{2\pi\epsilon l}{\ln(r_s/r_1)} \qquad (4\text{-}2)$$

where l is the length of the shield.

For applications such as metal detectors or radio direction finders, or between most power-transformer windings, it is desirable to shield against electric fields but not magnetic fields. The shielding technique discussed in this section provides this effect. To avoid affecting magnetic fields, a shield must be grounded at only one point. It must contain no closed loops or large surfaces through which eddy currents can flow. Such a shield is called a *Faraday* shield. Most applications, however, require shielding against both electric and magnetic fields. We will discuss magnetic fields in subsequent sections.

4-2 REDUCTION OF INDUCTIVE (MAGNETIC) COUPLING

Inductive coupling is much more difficult to reduce than capacitive coupling, since there is no magnetic equivalent of a conductive shield. A ferromagnetic sleeve would confine the stray magnetic field, as desired, but would work well only at low frequencies. At higher frequencies, the permeability of ferromagnetic materials decreases, thus reducing their effectiveness. A non-magnetic conductive shield, however, can reduce magnetic coupling if properly used. The technique is to generate a current flowing lengthwise in the shield such that its magnetic field cancels the undesired noise field. To provide a path for this current, some type of connection to each end of the shield is obviously necessary. The shield may be connected to any'of several points, however, and we will investigate the advantages of each choice.

The varying magnetic field due to the alternating current in the center lead induces a voltage into the shield, as it would into any nearby lead. The shield behaves as the secondary of a transformer, with the center lead acting as the primary. Let us now bend the cable into a circle and connect the opposite ends of the shield to each other but to nothing else. The shield then forms a new current loop shaped like a toroid (doughnut), which behaves as the shorted secondary of a transformer. The voltage induced into the shield loop produces a current in the loop that flows in the direction opposite to the current in the center lead. This current, in turn, generates a magnetic field outside the shield. Its direction is opposite to the original field and subtracts from it, as shown in the cross-sectional view in Figure 4-5. There, the symbols $\otimes$ and $\odot$ represent current flowing into and out of the page, respectively. We will later show that, if the current is uniformly distributed around the shield, it produces no additional magnetic field inside the shield. We also shall observe that the field outside the shield is just as it would be if the shield current were flowing in the center conductor.

The shield current increases, and the net magnetic field outside the shield decreases, until they reach equilibrium. At this point, the total voltage drop in

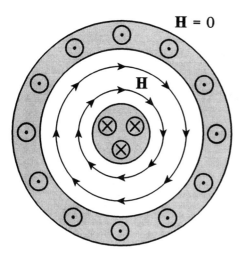

Figure 4-5 Currents and magnetic fields in shielded cable

the shield loop is equal to the voltage that the remaining magnetic field induces into the shield. The voltage drop in the shield is due to its resistance and leakage inductance. It would be zero if the shield were a perfect conductor and were perfectly coupled to the center lead. The net magnetic field outside the shield would then be zero, and no noise could be inductively coupled onto any other leads outside the shield.

In practice, the opposite ends of a shielded cable are usually far apart, and we cannot bend the cable into a circle as described above. If we could do this, we would not need the cable at all. Since we cannot, a non-inductive connection between the shield ends cannot then be realized, even approximately. Even if it could, the current flowing in the shield would have to return via the connecting lead. This current would generate still another magnetic noise field outside the shield, so the total field would not be reduced. Therefore, connecting the opposite ends of a shield by a wire is *not* an effective method of inductive noise suppression.

An effective method, however, is somehow to force the current flowing on the center conductor to return via the shield. We shall now investigate means to achieve this goal. The circuit in Figure 4-6 depicts two leads, designated 1 and 2, with a tubular shield around lead 1. The transformer analogy mentioned earlier is plainly visible in the figure. The shield is grounded at both ends. All stray inductances are shown, and it is important to understand the physical meaning of each inductance. Again we recall that no inductance is meaningful unless it forms part of a closed current loop. Inductances L_1, L_2, and L_{12} are as defined in Chapter 3, and their associated current loops are again loops 1 and 2. Remember that L_G forms a part of each of these three inductances; therefore, it does not appear separately in the figure.

The shield and the ground return together form the shield loop, mentioned earlier, which carries the shield current I_s. Inductance L_s is the flux generated by I_s, divided by the current I_s. It may be calculated in the same way as L_1 except that the

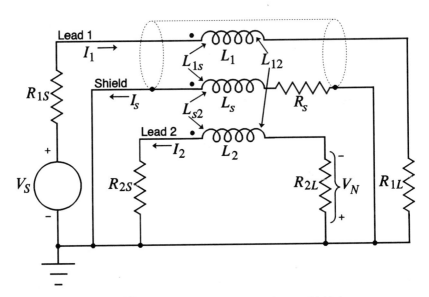

Figure 4-6 Inductances in two signal paths, one shielded

shield behaves like a thicker wire that replaces lead 1 in this calculation. Similarly, the mutual inductance L_{s2} between the shield and lead 2 is the flux generated by I_s that also links loop 2 divided by I_s. Loop 2 comprises lead 2 and the ground return. Inductance L_{s2} may be calculated like L_{12}, with the shield again behaving as a thicker wire. Only L_{1s} remains, which we will calculate later.

The loop equations for the circuit of Figure 4-6 are

$$(R_1 + j\omega L_1)I_1 - j\omega L_{1s}I_s - j\omega L_{12}I_2 = V_S \qquad (4\text{-}3)$$

$$-j\omega L_{1s}I_1 + (R_s + j\omega L_s)I_s + j\omega L_{s2}I_2 = 0 \qquad (4\text{-}4)$$

$$-j\omega L_{12}I_1 + j\omega L_{s2}I_s + (R_s + j\omega L_2)I_2 = 0 \qquad (4\text{-}5)$$

where $R_1 = R_{1S} + R_{1L}$ and $R_2 = R_{2S} + R_{2L}$ as in Chapter 3. The circuit of Figure 4-7 is equivalent to that of Figure 4-6, as we may show by writing its loop equations and comparing them to Equations (4-3) through (4-5). We will show that some of the equivalent inductances in Figure 4-7 are leakage inductances that are zero under ideal conditions, which will simplify the equivalent circuit. In a practical shielded cable, these inductances will be small but not zero. To analyze the effects of leakage inductance, we must include them. Their effect may be easier to visualize in Figure 4-7 than in Figure 4-6, since there is no coupling between any of the equivalent inductances shown in Figure 4-7.

We shall now calculate L_{1s}. Consider the current I_s flowing through a shield with no center conductor, as shown in Figure 4-8. By Ampere's law, for any field $\mathbf{H}_i$ inside the shield,

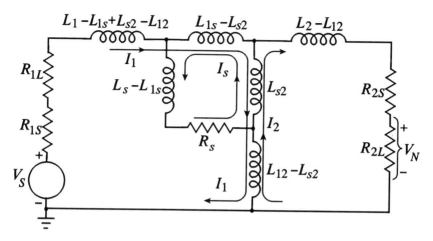

Figure 4-7 Equivalent circuit of shield and lead inductances

$$\oint \mathbf{H}_i \cdot dl = 0 \tag{4-6}$$

since no current can flow inside any closed path of integration inside the shield. However, by symmetry, any field would have to be tangential. It also would have to be constant over a path of constant radius. Thus, for any circular path inside the shield and concentric with its center,

$$\oint \mathbf{H}_i dl = 2\pi r |\mathbf{H}_i| = 0 \tag{4-7}$$

Therefore, $\mathbf{H} = 0$ everywhere inside the shield, and thus the entire field produced by I_s lies outside the shield.

If we now add a conductor within the shield, any flux Φ_s due to I_s also encircles the inner conductor and thus links loop 1. The shield inductance, or the inductance of the loop containing the current I_s, is

$$L_s = \frac{\Phi_s}{I_s} \tag{4-8}$$

The mutual inductance between the conductor and the shield is defined to be the flux Φ_1 linking loop 1 divided by I_s:

$$L_{1s} = \frac{\Phi_1}{I_s} = \frac{\Phi_s}{I_s} = L_s \tag{4-9}$$

Therefore, for a *uniform* shield current, the mutual inductance L_{1s} between the center conductor and its shield equals the self inductance L_s of the shield. Note, however, that there is some magnetic flux completely inside the shield due to I_1, which links loop 1 but not the shield loop. That flux causes the inductance L_1 of the

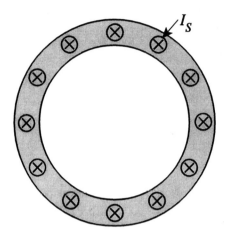

Figure 4-8   Current in hollow tubular shield

inner lead to be greater than L_s and thus $L_1 L_s > L_s^2 = L_{1s}^2$. Since $L_1 L_s > L_{1s}^2$, the center conductor and the shield are *not* perfectly coupled.

We will now prove another important result. The magnetic field outside the shield due to I_s appears as if this current were flowing in a line located approximately on its axis. Thus, if I_1 is equal to I_s and is flowing in the same direction, the currents produce identical fields everywhere outside the shield. Let Φ_{12} and Φ_{s2} represent the flux linking loop 2 due to I_1 and I_s, respectively. Then $\Phi_{12} = \Phi_{s2}$ since loop 2 is completely outside the shield where the fields must be identical. Thus, from the definitions of L_{s2} and L_{12}, it follows that

$$L_{12} = \frac{\Phi_{12}}{I_1} = \frac{\Phi_{s2}}{I_s} = L_{s2} \qquad \textbf{(4-10)}$$

If we calculate L_{s2} and L_{12} using the methods of Chapter 3, they will always turn out to be equal.

These equalities are exact only if the shield encases loop 1 for its entire closed path, like a hose with its ends coupled. Of course, this cannot be true in a real circuit, since connections must be made to the cable in order for it to be useful. As with capacitive shielding, we must keep these pigtail leads as short as possible. If they are short enough so that $L_{1s} = L_s$ and $L_{s2} = L_{12}$, as shown above, then Figure 4-7 reduces to Figure 4-9. If R_s increases until it becomes an open circuit, then Figure 4-9 becomes identical with Figure 3-11, which is the same circuit without a shield.

To find V_N we first derive an expression for I_2. We may substitute $L_{1s} = L_s$ and $L_{s2} = L_{12}$ into Equations (4-3) through (4-5), or we may write the desired equations directly from Figure 4-9. Solving the equations for I_2 yields

$$I_2 = \cfrac{j\omega L_{12} R_s V_S}{\begin{aligned} &(j\omega)^3 (L_1 - L_s)(L_s L_2 - L_{12}^2) + \\ &(j\omega)^2 [(L_1 L_s - L_s^2) R_2 + (L_1 L_2 - L_{12}^2) R_s + (L_s L_2 - L_{12}^2) R_1] + \\ &j\omega(L_1 R_s R_2 + L_s R_1 R_2 + L_2 R_1 R_s) + R_1 R_s R_2 \end{aligned}} \qquad \textbf{(4-11)}$$

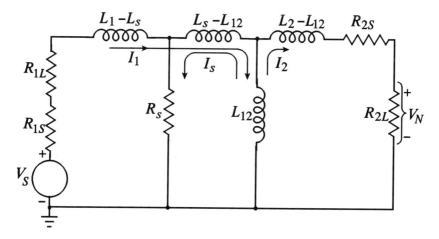

Figure 4-9   Shield and lead inductances without pigtails

Since $V_N = I_2 R_{2L}$, the expression for V_N is

$$V_N = \frac{j\omega L_{12} R_s R_{2L} V_S}{\begin{aligned}(j\omega)^3 (L_1 - L_s)(L_s L_2 - L_{12}^2) + \\ (j\omega)^2 [(L_1 L_s - L_s^2) R_2 + (L_1 L_2 - L_{12}^2) R_s + (L_s L_2 - L_{12}^2) R_1] + \\ j\omega (L_1 R_s R_2 + L_s R_1 R_2 + L_2 R_1 R_s) + R_1 R_s R_2\end{aligned}}$$

(4-12)

If we take the limit of Equation (4-12) as R_s approaches infinity, then it reduces to Equation (3-45), as expected. On the other hand, if R_s is zero, then I_2 and V_N are zero, so there is no coupled noise. Then there can be no voltage across the inductors labeled L_{12} and $L_s - L_{12}$, and therefore no current through them. Thus, if $R_s = 0$, I_s must be equal to I_1, which is the ideal case.

The transfer function of Equation (4-12) has three poles, with a zero at the origin. Solving the cubic equation would let us find the locations of the poles for specific values of inductance and resistance. The general formulas, however, are too cumbersome for practical use. The pole locations are of little value anyhow, since the function can be plotted without this information.

For example, let us assume the values in the example of Chapter 3, for which $R_1 = R_2 = 80\ \Omega$, $L_{12} = 1.9\ \mu H$, and $L_1 = L_2 = 2.76\ \mu H$. Here we also assume $L_s = 2.62\ \mu H$, corresponding to a shield diameter of 0.02 in. This is smaller than most shields, but it allows us to use all of the lead spacings presented in Chapter 3, including the close spacing of 0.05 in. The plot of $|V_N/V_S|$ versus frequency, for various values of R_s, appears in Figure 4-10. The topmost curve corresponds to no shield at all and is equivalent to the top curve in Figure 3-12, although the vertical scale is different.

Now assume that the shield is on lead 2 instead of lead 1, as shown in Figure 4-11. Equations (4-3) through (4-5) still apply, since all voltages, currents,

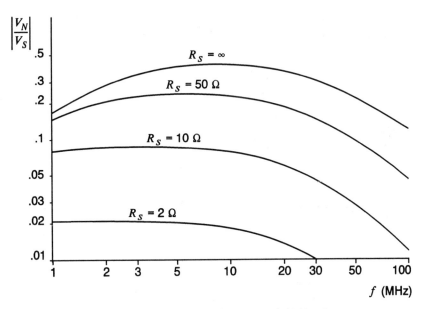

Figure 4-10   Induced noise vs. frequency and shield resistance

inductances, and resistances are similarly related. The inductance *values,* however, will usually be different. Here the equalities $L_s = L_{s2}$ and $L_{12} = L_{1s}$ replace Equations (4-9) and (4-10). Nevertheless, the expression for V_N will turn out to be the same as Equation (4-12).

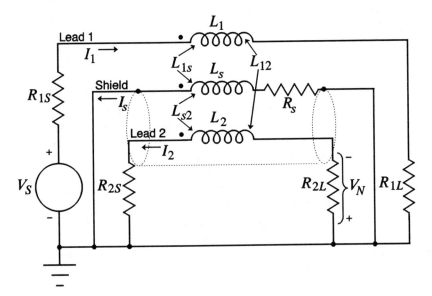

Figure 4-11   Inductances in two signal paths, path 2 shielded

4-3 EFFECTS OF SHIELD RESISTANCE

From Figure 4-10 we note that the shield resistance R_s has a critical effect on the coupled noise magnitude. It is shunted by the series combination of the two inductances L_s–L_{12} and L_{12}, which together equal L_s. At high frequencies this impedance is much greater than R_s, and R_s passes most of the current. Then $I_s \approx I_1$, and very little noise current reaches the load. For these frequencies the shield greatly reduces V_N. Conversely, at low frequencies, the inductance passes most of the current and $I_s \ll I_1$. Then the shield has little effect, since little current is flowing through it and R_s behaves like an open circuit. The circuit of Figure 4-7 then reduces to Figure 3-11, which contains no shield.

To establish a quantitative measure of the value of a shield, we define its *shielding effectiveness,* designated S. It is the ratio of the value of V_N without the shield, to its value with the shield, and is usually expressed in decibels. The value of S depends on the frequency and on all inductances and resistances, and is given by

$$S = 20 \log \frac{|V_N|_{no\ shield}}{|V_N|_{shield}}$$

$$= 20 \log \left| \frac{\begin{array}{c}(j\omega)^3(L_1 - L_s)(L_sL_2 - L_{12}^2) + \\ (j\omega)^2[(L_1L_s - L_s^2)R_2 + (L_1L_2 - L_{12}^2)R_s + (L_sL_2 - L_{12}^2)R_1] + \\ j\omega(L_1R_sR_2 + L_sR_1R_2 + L_2R_1R_s) + R_1R_sR_2\end{array}}{|R_s[(j\omega)^2(L_1L_2 - L_{12}^2) + j\omega(L_2R_1 + L_1R_2) + R_1R_2]|} \right|$$

$$= 20 \log \left| 1 + \frac{\begin{array}{c}(j\omega)^3(L_1 - L_s)(L_sL_2 - L_{12}^2) + \\ (j\omega)^2[(L_1L_s - L_s^2)R_2 + (L_sL_2 - L_{12}^2)R_1] + j\omega L_sR_1R_2\end{array}}{(j\omega)^2(L_1L_2 - L_{12}^2)R_s + j\omega(L_2R_1R_s + L_1R_2R_s) + R_1R_2R_s} \right| \quad \textbf{(4-13)}$$

The argument of the logarithm in Equation (4-13) has three zeroes and two poles, none of which are at the origin. At low frequencies it approaches the constant value of 1, its logarithm is 0, and the shield provides no attenuation. At high frequencies the argument becomes proportional to frequency, and the shielding effectiveness increases 20 dB per decade.

Since the scales in Figure 4-10 are logarithmic, the shielding effectiveness may be read directly from that graph if it is properly rescaled. It is proportional to the distance between the topmost curve, which signifies no shield, and the curve corresponding to the actual shield resistance. It is also plotted directly, for the inductance and resistance values assumed above, in Figure 4-12.

Due to the complexity of Equation (4-13), we normally assume a simpler case when comparing the effectiveness of various shields. We replace the voltage source

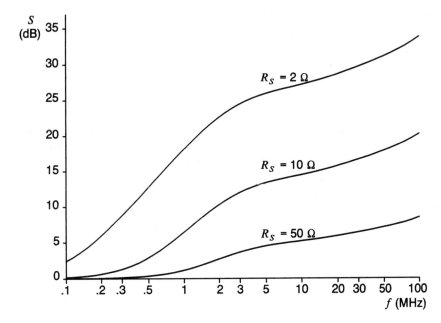

Figure 4-12   Shielding effectiveness vs. frequency and resistance

with a current source and assume the load impedance to be infinite. This causes both R_1 and R_2 to appear infinite, and Equation (4-13) reduces to

$$S = 20 \log \left| 1 + \frac{j\omega L_s}{R_s} \right| \tag{4-14}$$

A graph of this function, for the same values of L and R_s, appears in Figure 4-13. This function has no poles and only one zero, at $\omega_c = R_s/L_s$. The frequency of this zero, $f_c = \omega_c/(2\pi) = R_s/(2\pi L_s)$, is known as the *shield cutoff frequency*. This cutoff frequency is visible in Figure 4-13 for each value of R_s, at the point where the respective diagonal asymptote crosses the horizontal axis. Below this frequency the shield is ineffective. The shield effectiveness is expressible in terms of f_c or ω_c by substituting for R_s/L_s:

$$S = 20 \log \left| 1 + j\frac{\omega}{\omega_c} \right| = 20 \log \left| 1 + j\frac{f}{f_c} \right| \tag{4-15}$$

For example, for the shield to achieve a 30-dB reduction in coupled noise,

$$20 \log \left| 1 + \frac{j\omega L_s}{R_s} \right| = 30$$

$$\left| 1 + \frac{j\omega L_s}{R_s} \right| = 10^{3/2} = \sqrt{1000} \tag{4-16}$$

$$\sqrt{1 + \left(\frac{\omega L_s}{R_s}\right)^2} = \sqrt{1000}$$

$$\frac{\omega L_s}{R_s} = \sqrt{1000 - 1} = 31.6$$

(4-16)
(cont.)

Therefore, the shield provides at least 30 dB of attenuation for all frequencies such that $\omega \geq 31.6\ \omega_c$ or $f \geq 31.6\ f_c$.

We must emphasize that this calculation applies only with infinite load impedances in both loops, which is unrealistic. With practical load impedances, the shielding effectiveness will be reduced, as is evident by comparing Figures 4-12 and 4-13. The simpler expression in Figure 4-13, however, often yields a reasonable estimate of any improvement obtained when we vary one or more circuit parameters.

4-4 CAUSES AND EFFECTS OF LEAKAGE FLUX

The preceding analysis assumes that $L_{1s} = L_s$ and $L_{12} = L_{s2}$, which we proved under ideal conditions. We shall now see that these equalities are not true if we use pigtail leads to connect the shield. A finite area, visible in Figure 4-14, exists

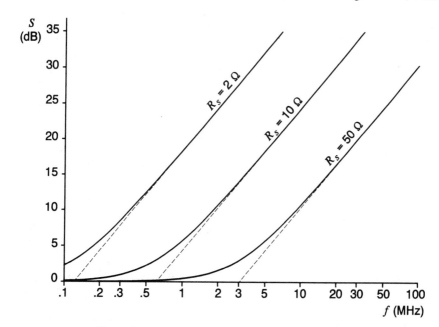

Figure 4-13 Shielding effectiveness for high impedances

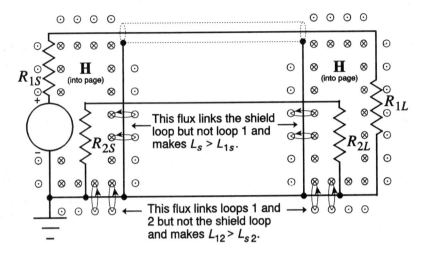

Figure 4-14   Leakage flux caused by pigtail leads

between the center lead and each ground connection. Any magnetic flux that passes through either of these areas between the leads must link loop 1 or the shield loop, but it cannot link both loops. If it links the shield loop, then since it cannot link loop 1, it increases L_s but not L_{1s} and causes them to become unequal. If this flux also links loop 2, it adds to L_{12} or L_{1s} but not both, and similarly causes them to become unequal. Without these equalities, we must use Figure 4-7 instead of Figure 4-9. With the additional nonzero inductances, V_N can be significant even if R_s is zero. It is important, therefore, to minimize the areas between the pigtail leads by using the shortest possible leads. A coaxial connector is even better, if mounted such that the shield contacts the ground for the full 360 degrees around the center lead.

We have already seen that, ideally, $I_s = I_1$ so that I_2 and V_N are zero. The current I_2 is nonzero only if some voltage is induced in series with that current loop. If the shield current I_s encounters any impedance while traversing the length of the shield, then a voltage does appear. It diverts some shield current into loop 2. Obviously, the shield resistance R_s is a portion of this impedance. We will use circuit theory to define the impedance more precisely.

The voltage induced into loop 2 may be found by opening both loop 2 and the ground return connection. Then the only path for I_1 to return is via the shield. A voltage V_s then appears between the opposite ends of the shield, and this voltage would be induced into loop 2 if the loop were still present.

We noted earlier that the shielded cable is effectively a transformer. It appears in Figure 4-15 as a four-node network. Resistance R_{1L}, of course, is not part of the shield, but it is included to emphasize that I_1 must return via the shield and therefore $I_s = I_1$.

When a current source I is applied between two nodes m and n in a network, it usually causes a voltage V to appear between any two nodes p and q in the network.

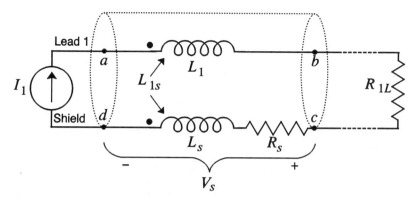

Figure 4-15 Transformer equivalent of a shielded source lead

The ratio of the voltage V to its corresponding current I is the transfer impedance between node pair mn and node pair pq. A transfer impedance can be calculated between any two node pairs, but the transfer impedance between pair cd and pair ad in Equation (4-16) is of special interest here. It relates the voltage V_s to the current I_1 and is called the *shield transfer impedance*, designated Z_s. It is given by

$$Z_s = \frac{V_{cd}}{I_{ad}} = \frac{V_s}{I_1} = \frac{-I_1 j\omega L_{1s} + I_s(R_s + j\omega L_s)}{I_1} = R_s + j\omega(L_s - L_{1s}) \qquad \textbf{(4-17)}$$

Thus, $V_s = I_1 Z_s$, and Z_s includes R_s and any leakage inductance $L_s - L_{1s}$.

Using Z_s, we may express V_N in Figure 4-6 (or Figure 4-7) for the more general case where $L_s \geq L_{1s}$, although we still assume $L_{s2} = L_{12}$. To keep the expression simple, we also still assume infinite load impedances, so that I_2 is zero and I_1 is a constant current source. Then from Equation (4-4),

$$-j\omega L_{1s} I_1 + (R_s + j\omega L_s)I_s = 0$$

$$I_s = I_1 \frac{j\omega L_{1s}}{R_s + j\omega L_s}$$

$$= I_1 \frac{j\omega L_{1s}}{R_s + j\omega(L_s - L_{1s}) + j\omega L_{1s}}$$

$$= I_1 \frac{j\omega L_{1s}}{Z_s + j\omega L_{1s}} \qquad \textbf{(4-18)}$$

With R_{2L} infinite, the loop equation for loop 2 becomes

$$-j\omega L_{12} I_1 + j\omega L_{s2} I_s + V_N = 0 \qquad \textbf{(4-19)}$$

Combining Equations (4-18) and (4-19) while still assuming $L_{12} = L_{s2}$ gives V_N in terms of I_1. We have

$$V_N = j\omega L_{12}(I_1 - I_s)$$

$$= j\omega L_{12}\left(I_1 - I_1\frac{j\omega L_{1s}}{Z_s + j\omega L_{1s}}\right)$$

$$= j\omega L_{12}I_1\left(\frac{Z_s}{Z_s + j\omega L_{1s}}\right) \tag{4-20}$$

If there were no shield, I_s and I_2 would both be zero and the equation for loop 2 would simply be

$$-j\omega L_{12}I_1 + V_N = 0$$

$$V_N = j\omega L_{12}I_1 \tag{4-21}$$

The shielding effectiveness for this case is the ratio of Equation (4-21) to Equation (4-20), expressed in decibels:

$$S = 20\log\frac{|V_N|_{no\ shield}}{|V_N|_{shield}}$$

$$= 20\log\left|\frac{j\omega L_{12}I_1}{j\omega L_{12}I_1\left(\dfrac{Z_s}{Z_s + j\omega L_{1s}}\right)}\right|$$

$$= 20\log\left|1 + \frac{j\omega L_{1s}}{Z_s}\right| \tag{4-22}$$

As ω decreases, ωL_{1s} decreases faster than $|Z_s|$, so S decreases and the shield eventually becomes ineffective. If Z_s is purely resistive (no leakage inductance), the problem equates to the shield cut off frequency issue discussed earlier. Treating shielding effectiveness in the manner shown here, however, is more general and allows R_s, L_s, and L_{1s} to vary with frequency in any way at all. This occurs at higher frequencies, due to skin effect, which we will discuss in Chapter 9. The shield transfer impedance then becomes a more complicated function of frequency,[1] and, for a braided shield, depends on the density of the braid. Nevertheless, if we know Z_s and L_{1s}, we can still use the above technique, including Equation (4-22).

The shield transfer impedance per unit length, designated Z_l, is defined as follows:

$$Z_l = \frac{dZ_s}{dl} = \frac{1}{I_s}\cdot\frac{dV_1}{dl} \tag{4-23}$$

It is defined as a derivative to allow for the possibility that it may vary along the length of the cable.

---

1. Edward F. Vance, *Coupling to Shielded Cables* (New York, N.Y.: John Wiley & Sons, Inc., 1978), pp. 37, 112–17, 131–47.

If the shield is on lead 2 instead of lead 1, as shown in Figure 4-16, the shield transfer impedance applies in the reverse direction. Here the magnetic flux from loop 1 induces a voltage into the shield that causes a current I_s to flow in it. In turn, I_s causes a voltage V_2, equal to $I_s Z_s$, to appear between nodes b and c, as shown. When the load is connected across these nodes, it will draw current and reduce this voltage, but a portion of V_2 will still appear across the load. This voltage is V_N, defined earlier.

We shall now investigate shielding effectiveness at frequencies for which Z_s is purely inductive. Again we keep the expression simple by assuming infinite load impedances, so that I_2 is zero and I_1 is a constant current source. Using the original circuit of Figure 4-6, we arrange its leads into the circular geometric pattern shown in Figure 4-17. The shielded cable and the ground return lead form a large circle of radius r_1, and the termination leads form two smaller circles, each of radius r_p. Loop 2 does not appear in Figure 4-17 since, with $I_2 = 0$, it does not influence the shielding effectiveness. For this geometry, the inductance of each circle is simple to calculate if we know the wire diameter d. From this we can find the inductances L_s and L_{1s} and substitute them into Equation (4-22).

We first compute L_1, setting I_s equal to zero. Loop 1 consists of the center conductor of the shielded cable, the outer halves of the termination loops, and the ground return lead. These leads form a nearly perfect circle. Inductance L_1 is the flux linking loop 1, divided by I_1. This is approximately equal to the inductance of a perfectly circular loop, for which the geometric formula[2] is

$$L_1 = \mu r_1 \left(\ln \frac{16 r_1}{d} - 2 \right) \tag{4-24}$$

We neglect the aberration caused by the outer halves of the small loops. We compute L_s similarly, with I_1 set to zero. The shield loop consists of the shield, the inner

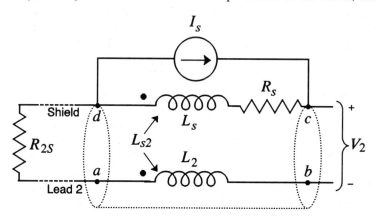

Figure 4-16 Transformer equivalent of a shielded sense lead

2. Simon Ramo, John R. Whinnery, and Theodore Van Duzer, *Fields and Waves in Modern Radio* (New York, N.Y.: John Wiley & Sons, Inc., 1984), p. 190.

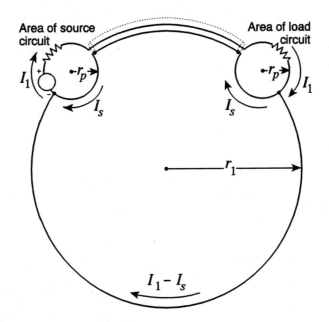

Figure 4-17   Circular lead pattern used to estimate typical stray inductances

halves of the termination loops, and the ground return lead. Inductance L_s is nearly equal to L_1, neglecting the aberration of the inner halves of the small loops. These approximations are adequate for L_1 and L_s when they appear by themselves in a formula.

To obtain L_{1s} we must examine the location of the flux due to I_1 and I_s. Let Φ_s represent the flux generated by I_s, which must, of course, link the shield loop. All of this flux also links loop 1 *except* the flux that encircles the pigtail leads (the inner halves of the two termination loops). Let Φ_{1s} represent the flux due to I_s that links loop 1, and let Φ_{ps} be the flux due to I_s that links the pigtail leads. Then

$$\Phi_s = \Phi_{1s} + \Phi_{ps} \qquad (4\text{-}25)$$

Now we set I_s equal to I_1. Since these two currents flow in opposite directions, they cancel each other in the ground return lead. They also produce equal and opposite fields outside the shielded cable, since the field caused by the shield current is just as if the current were flowing in its center conductor. Thus, the only magnetic field outside the shield is near the two termination loops. This field is the same as what would be generated by the current I_1 or I_s flowing around each termination loop. Let Φ_p be the total flux linking the termination loops with $I_s = I_1$. The outer half of Φ_p is due to I_1, whereas the inner half is due to I_s and is therefore Φ_{ps}. Thus,

$$\Phi_{ps} = \frac{\Phi_p}{2} \qquad (4\text{-}26)$$

By the definition of mutual inductance,

$$L_{1s} = \frac{\Phi_{1s}}{I_s} = \frac{\Phi_s - \Phi_{ps}}{I_s} = \frac{\Phi_s - \Phi_p/2}{I_s} = L_s - \frac{L_p}{2} \tag{4-27}$$

where L_p is the combined self inductance of the two termination loops. They are effectively connected in series, and we assume that there is no mutual inductance between them. Their combined inductance L_p is therefore

$$L_p = 2\mu r_p \left(\ln \frac{16 r_p}{d} - 2 \right) \tag{4-28}$$

By setting these loop sizes equal to the approximate sizes of the (non-circular) loops in a real circuit, we can estimate the inductances in that circuit. Typical dimensions might be $r_1 = 0.5$ m, $r_p = 5$ cm, and $d = 0.64$ mm (#22 AWG). For such a circuit, from Equation (4-24),

$$L_1 \approx L_s \approx 4\pi \cdot 10^{-7} \cdot 0.5 \left(\ln \frac{16 \cdot 0.5}{0.64 \cdot 10^{-3}} - 2 \right) \text{H}$$

$$= 62.83 \cdot 10^{-8} \, (\ln 12500 - 2) \text{ H}$$

$$= 4.67 \, \mu\text{H} \tag{4-29}$$

From Equation (4-28),

$$L_p = 2 \cdot 4\pi \cdot 10^{-7} \cdot 0.05 \left(\ln \frac{16 \cdot 0.05}{0.64 \cdot 10^{-3}} - 2 \right) \text{H}$$

$$= 125.66 \cdot 10^{-9} \, (\ln 1250 - 2) \text{ H}$$

$$= 125.66 \cdot 10^{-9} \cdot 5.13 \text{ H}$$

$$= 0.64 \, \mu\text{H} \tag{4-30}$$

and, from Equation (4-27),

$$L_{1s} = L_s - \frac{L_p}{2} = \left(4.67 - \frac{0.64}{2} \right) \mu\text{H} = 4.35 \, \mu\text{H} \tag{4-31}$$

Knowing these inductances, we can calculate the shielding effectiveness. We assume perfect conductivity, so $R_s = 0$.

$$S = 20 \log \left| 1 + \frac{j\omega L_{1s}}{Z_s} \right|$$

$$= 20 \log \left| 1 + \frac{j\omega L_{1s}}{R_s + j\omega(L_s - L_{1s})} \right|$$

$$= 20 \log \left(1 + \frac{L_{1s}}{L_s - L_{1s}} \right)$$

$$= 20 \log \left(1 + \frac{4.35}{0.32} \right) = 23.2 \text{ dB} \tag{4-32}$$

Since the shield resistance is zero, there is no cutoff frequency and, moreover, the shield effectiveness does not depend on frequency. Nevertheless, the attenuation of 23.2 dB is far less than what we might have expected from a solid shield of zero resistance. To devise an effective shield against magnetic coupling requires additional effort.

4-5 STRAY CURRENT RETURN PATHS

The poor attenuation of the shield discussed above occurs because the shield current I_s is not exactly equal to the current I_1 on the center lead. The shield transfer impedance causes a portion of I_1 to return via the common ground connection instead of the shield. In the example given in Section 4-4, we can easily calculate the ratio of these currents. From Equation (4-4), with $I_2 = 0$ since we assumed R_{2L} to be infinite,

$$\frac{I_s}{I_1} = \frac{j\omega L_{1s}}{R_s + j\omega L_s} = \frac{L_{1s}}{L_s} = \frac{4.35}{4.67} = 0.931 \qquad \textbf{(4-33)}$$

Thus, in this example, only 93.1 percent of the center conductor current will return via the shield. The remaining 6.1 percent will return via the common ground and cause noise to be coupled into loop 2.

The impedance in the shield return path is Z_s, and the impedance in the ground return path is $j\omega L_{1s}$. Although $j\omega L_{1s}$ is normally much greater than $|Z_s|$, their ratio is not infinite, and the currents will divide according to this ratio. If $|Z_s|$ is already as small as possible, the only other possibility is to increase $j\omega L_{1s}$.

It is easier to understand the relationships of these currents to the coupled noise if we define two new current loops. Let $I_d = (I_1 + I_s)/2$, and let $I_c = I_1 - I_s$. The current I_d is called the *differential-mode* or *signal* current. It flows down the center conductor and returns via the shield without flowing in the ground return. The current I_c is called the *common-mode* current[3]. This current flows through the common ground and splits equally between the shield and the center lead. Both currents appear in Figure 4-18. If $I_1 = I_s$, as desired, then $I_c = 0$; therefore, any common-mode current is undesirable.

With I_d and I_c defined as above, $I_1 = I_d + \frac{1}{2} I_c$ and $I_s = I_d - \frac{1}{2} I_c$. Substituting these expressions into Equations (4-3) and (4-4) yields

$$(R_1 + j\omega L_1)(I_d + \frac{1}{2} I_c) - j\omega L_{1s}(I_d - \frac{1}{2} I_c) - j\omega L_{12} I_2 = V_S$$

$$[R_1 + j\omega(L_1 - L_{1s})]I_d + \frac{1}{2}[R_1 + j\omega(L_1 + L_{1s})]I_c - j\omega L_{12} I_2 = V_S \qquad \textbf{(4-34)}$$

$$-j\omega L_{1s}(I_d + \frac{1}{2} I_c) + (R_s + j\omega L_s)(I_d - \frac{1}{2} I_c) + j\omega L_{s2} I_2 = 0$$

$$[R_s + j\omega(L_s - L_{1s})]I_d - \frac{1}{2}[R_s + j\omega(L_s + L_{1s})]I_c + j\omega L_{s12} I_2 = 0 \qquad \textbf{(4-35)}$$

---

3. In telephony, differential-mode and common-mode currents are commonly known as metallic and longitudinal currents, respectively.

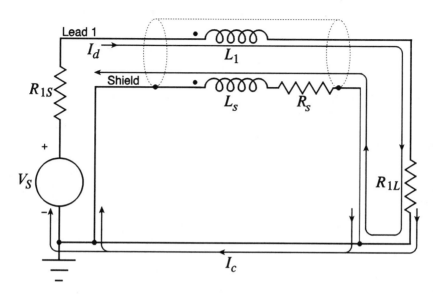

Figure 4-18 Common-mode and differential-mode currents

Adding Equations (4-34) and (4-35) then gives the loop equation for the differential-mode current loop I_d. Subtracting Equation (4-35) from Equation (4-34) gives the equation for the common-mode current loop I_c. Substituting for I_1 and I_s in Equation (4-5) gives the loop equation for I_2 in terms of I_d and I_c. These three equations, shown below, form a new set of simultaneous equations for this circuit.

$$[R_1+R_s+j\omega(L_1+L_s-2L_{1s})]I_d+ \tfrac{1}{2}[R_1-R_s+j\omega(L_1-L_s)]I_c-j\omega(L_{12}-L_{s2})I_2=V_S$$
$$\text{(4-36)}$$

$$[R_1-R_s+j\omega(L_1-L_s)]I_d+ \tfrac{1}{2}[R_1+R_s+j\omega(L_1+L_s+2L_{1s})]I_c-j\omega(L_{12}+L_{s2})I_2=V_S$$
$$\text{(4-37)}$$

$$-j\omega(L_{12}-L_{s2})I_d- \tfrac{1}{2}j\omega(L_{12}+L_{s2})I_c+(R_2+j\omega L_2)I_2=0 \qquad \text{(4-38)}$$

The self impedance of the differential-mode current loop is called the *differential-mode impedance*. From Equation (4-36), it is equal to $R_1+R_s+j\omega(L_1+L_s-2L_{1s})$, or equivalently, $R_1+j\omega(L_1-L_{1s})+Z_s$. Since I_d is the desired current path, this impedance should be low, to have minimum effect. The self impedance of the common-mode current loop is called the *common-mode impedance*, which should be as high as possible since this current is undesirable. From Equation (4-37), the common-mode impedance is $\tfrac{1}{2}[R_1+R_s+j\omega(L_1+L_s+2L_{1s})]$.

From Equation (4-38) we can observe the effect on I_2, and therefore V_N, due to each current, I_d and I_c. In particular, if $L_{12}=L_{s2}$ as assumed earlier, then I_d has no effect at all on I_2 or V_N. Then all coupled noise is proportional to I_c. This observation directly illustrates the undesirability of the common-mode current I_c. We will now investigate various methods of reducing this current.

A very effective method of eliminating I_c is to remove the connection between R_{1L} and the ground return and connect R_{1L} only to the shield, thus opening the current loop. This circuit is shown in Figure 4-19. In effect, L_1, L_s, and L_{1s} become infinite, but the differences between them remain finite. Note that R_{2L} must still be connected to the original ground return and not to the shield. Otherwise, L_{12} and L_{s2} also would become infinite and we would not achieve the desired effect.

We will analyze the circuit in Figure 4-19 by setting $I_c = 0$ in Equations (4-36) and (4-38) and eliminating Equation (4-37) since that loop no longer exists. The equations become

$$[R_1 + R_s + j\omega(L_1 + L_s - 2L_{1s})]I_d - j\omega(L_{12} - L_{s2})I_2 = V_S \qquad \textbf{(4-39)}$$

$$-j\omega(L_{12} - L_{s2})I_d + (R_2 + j\omega L_2)I_2 = 0 \qquad \textbf{(4-40)}$$

We must again remember that inductances are meaningful only with respect to closed current loops. Each inductance, L_1, L_s, and L_{1s}, by itself is no longer a part of a closed loop and is therefore meaningless. The expression $L_1 + L_s - 2L_{1s}$, however, is the self inductance of the differential-mode current loop, and we may use it in the same way as any other inductance. Similarly, $-(L_{12} - L_{s2})$ is the mutual inductance between the differential-mode current loop and loop 2.

The solution of Equations (4-39) and (4-40) gives the following expressions for I_2 and V_N:

$$I_2 = \frac{j\omega(L_{s2} - L_{12})V_S}{[R_1 + R_s + j\omega(L_1 + L_s - 2L_{1s})](R_2 + j\omega L_2) - (j\omega)^2(L_{s2} - L_{12})^2} \qquad \textbf{(4-41)}$$

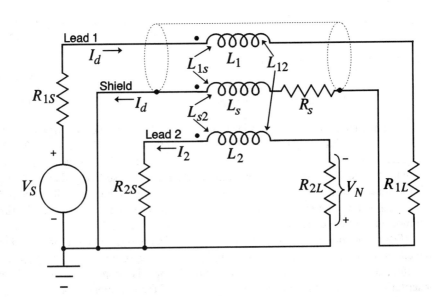

Figure 4-19   Isolation of signal return path

$$V_N = I_2 R_{2L}$$

$$= \frac{j\omega(L_{s2} - L_{12})R_{2L}V_S}{[R_1 + R_s + j\omega(L_1 + L_s - 2L_{1s})](R_2 + j\omega L_2) - (j\omega)^2(L_{s2} - L_{12})^2} =$$

$$\frac{j\omega(L_{s2} - L_{12})R_{2L}V_S}{(j\omega)^2[(L_1 + L_s - 2L_{1s})L_2 - (L_{s2} - L_{12})^2] + j\omega[(L_1 + L_s - 2L_{1s})R_2 + L_2(R_1 + R_s)] + (R_1 + R_s)R_2}$$

$$(4\text{-}42)$$

Unlike the earlier solution with R_{2L} connected to ground, this solution does not assume the equivalence of any inductances. As noted earlier, since $I_c = 0$, if $L_{12} = L_{s2}$ then there is no coupled noise at all, whatever the other inductances and resistances. All coupled noise is due to the difference between these two inductance values. The equality of L_{12} and L_{s2} depends on the quality of the shielded cable. The center conductor must coincide with the exact center of the shield, and the cross section of the shield must be a perfect circle. These requirements are most closely realized with RG/U type cable. Also, the conductor carrying I_2 must be far enough away from the shield so that the shield current distributes itself uniformly around the shield. If these objectives can be achieved, then the shielding effectiveness is limited only by the inductance of the pigtail leads, if present. Attenuations as high as 80 dB are common with this circuit[4].

The obvious difficulty with this technique is that it is impossible to ground the load resistance. If R_{1L} is the input circuit of an amplifier, it must be a differential amplifier. Otherwise, we must connect the amplifier ground to the shield instead of the chassis. Then the entire amplifier is referenced to the shield, and so is its output. We cannot reference the entire receiving circuit to the shield, for then R_{2L} also would have to be connected to the shield and the circuit would not be that of Figure 4-19. Therefore, the reference point must be changed at some point in the receiver, again requiring a differential amplifier. Such an amplifier responds to the voltage difference between its input terminals but not to the voltage between either input and ground. We can still reference its output, however, to its own ground as required. It thus rejects the common-mode current. We will discuss differential amplifiers later, with balanced circuits.

For the above reasons, it is often impossible to remove the ground connection between R_{1L} and ground. In such cases other methods must be used to reduce the common-mode current. These methods are discussed in the next section.

4-6 METHODS OF ADDING COMMON-MODE IMPEDANCE

To reduce the common-mode current, we must increase the common-mode impedance. We can increase this impedance, without affecting the differential-mode impedance, by increasing L_1, L_s, and L_{1s} by the same amount.

---

4. Henry W. Ott, *Noise Reduction Techniques in Electronic Systems* (New York, N.Y.: John Wiley & Sons, Inc., © 1988), pp. 56–59. Reprinted by permission of John Wiley & Sons, Inc.

Additional inductance may be inserted anywhere in loop 1 and the shield loop. To keep $L_1 + L_s - 2L_{1s}$ constant, we also must add an equal amount of mutual inductance between the loops. Therefore, to avoid increasing the differential-mode impedance, we must add equal inductances to the center conductor and the shield, and they must be perfectly coupled. If we insert an inductor L_g into a lead carrying both I_1 and I_s, as shown in Figure 4-20, it increases L_1, L_s, and L_{1s} by the value of L_g. It thus increases all of them equally, as required. Unfortunately, the only leads carrying both I_1 and I_s are the ground leads. To insert an inductor into either of these leads, either the source or load resistor must be ungrounded. If this can be done, then the highest possible inductance should be inserted at that point. An open circuit is effectively an infinite inductance and is the best option, as discussed earlier.

If neither load can be ungrounded, then the only leads where an inductor can be inserted will be carrying I_1 or I_s, but not both. Therefore, we must add two equal inductors to the center lead and the shield, respectively, and they must be as tightly coupled as possible. This device is called a *common-mode choke*, shown in Figure 4-21, where L_c is the self inductance of each inductor. The choke is normally wound on a ferrite core for optimum frequency response. Since two real coils cannot be perfectly coupled, the coupling coefficient k is used as a measure of the degree of coupling. For two equal inductors, k is the ratio of their mutual inductance to either self inductance, so the mutual inductance of the choke is kL_c. We may evaluate the effectiveness of the common-mode choke using the solution of Equations (4-3) through (4-5) as before, but with L_c added to L_1 and L_s, and with kL_c added to L_{1s}.

For example, let us add a 50 μH common-mode choke, with a coupling coefficient of 0.99, to the example from Section 4-4. Then L_s increases from

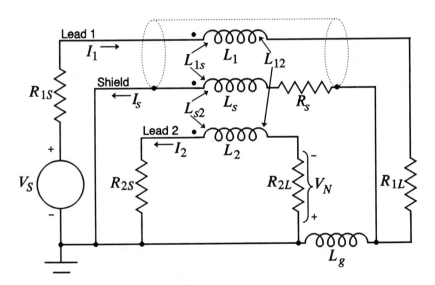

Figure 4-20   Intentional impedance in undesired signal return path

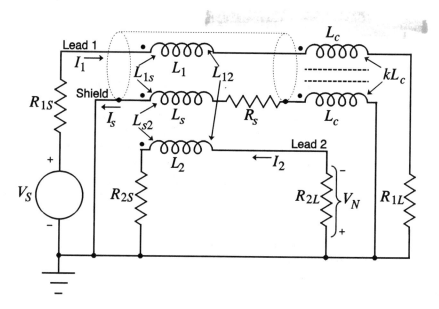

Figure 4-21 Common-mode choke in signal path

4.67 μH to 54.67 μH, and since $kL_c = 49.5$ μH, L_{1s} increases from 4.35 μH to 53.85 μH. Rewriting Equation (4-32) with these new inductance values yields the new shielding effectiveness:

$$S = 20 \log \left(1 + \frac{L_{1s}}{L_s - L_{1s}}\right) = 20 \log \left(1 + \frac{53.85}{0.82}\right) = 36.5 \text{ dB} \qquad \textbf{(4-43)}$$

This is a substantial improvement; however, the coupling coefficient must be very high (see Problem 11).

To achieve the required coupling coefficient, a common-mode choke may be made by winding a portion of the shielded cable onto a toroid core. Then, using arguments presented in Section 4-2, we may show that although the coupling is not perfect, the increase in L_{1s} is equal to the increase in L_s. This type of choke is very effective. At higher frequencies, a single turn through the toroid core may suffice.

With all common-mode chokes, there is danger of saturation due to direct currents flowing through the shield. These reduce the permeability of the core and reduce the inductance to a small fraction of its original value. If the circuit design is such that DC or low-frequency AC can flow through the winding of a common-mode choke, that current should be blocked with a capacitor. The capacitor must have low inductance, since this inductance adds to L_s but not L_{1s} and reduces the coupling.

Another concern is cross-talk between two signal circuits, each containing a common-mode choke. A common-mode choke is a transformer, and if two chokes are near each other, there may be some mutual inductance between them, where we do not want it. This is avoidable by separating the chokes, or by shielding them if separation is not possible.

4-7 BALANCED CIRCUITS

The strategy common to all methods of inductive noise reduction is to force I_s to equal I_1. Since we consistently assume $L_{s2} = L_{12}$, any coupled noise then cancels. With the circuits previously discussed, I_s was generated by the intentional inductive coupling L_{1s} between loop 1 and the shield loop. The current I_s would be most nearly equal to I_1 if the shield transfer impedance Z_s were minimized.

An alternate approach is to add a generator into the shield loop to generate I_s. Specifically, we move half the source voltage V_S to the shield loop and reverse its polarity. We intentionally add resistance so that R_s becomes equal to R_1, and we make L_s equal to L_1 without regard for L_{1s}. If L_{s2} is still equal to L_{12}, then, by symmetry, this guarantees $I_1 = I_s$. With this condition, the circuit is said to be *balanced*.

The loop equations for the balanced circuit are identical with Equations (4-3) through (4-5), except that the right members of Equations (4-3) and (4-4) are both $V_S/2$. This circuit is easiest to analyze in terms of the differential and common-mode currents I_d and I_c. Equations (4-36) through (4-38) become, for the balanced circuit,

$$[R_1 + R_s + j\omega(L_1 + L_s - 2L_{1s})]I_d + \frac{1}{2}[R_1 - R_s + j\omega(L_1 - L_s)]I_c - j\omega(L_{12} - L_{s2})I_2 = V_S \tag{4-44}$$

$$[R_1 - R_s + j\omega(L_1 - L_s)]I_d + \frac{1}{2}[R_1 + R_s + j\omega(L_1 + L_s + 2L_{1s})]I_c - j\omega(L_{12} + L_{s2})I_2 = 0 \tag{4-45}$$

$$-j\omega(L_{12} - L_{s2})I_d - \frac{1}{2}j\omega(L_{12} + L_{s2})I_c + (R_2 + j\omega L_2)I_2 = 0 \tag{4-46}$$

We now solve these equations for I_2 using Cramer's rule. The numerator of the solution is $-\frac{1}{2}V_S j\omega(L_{12} + L_{s2})[R_1 - R_s + j\omega(L_1 - L_s)]$, since we still assume $L_{12} - L_{s2}$ to be zero. Thus $|I_2|$ and $|V_N|$ are proportional to $|R_1 - R_s + j\omega(L_1 - L_s)|$, so minimizing $|R_1 - R_s|$ and $|L_1 - L_s|$ minimizes the noise.

To minimize $|L_1 - L_s|$ we must, of course, make $L_s = L_1$, which previously has not been necessary. Since there must be some magnetic flux in the space within the shield that links loop 1 but not the shield loop, this is impossible. However, although L_{s2} must still be equal to L_{12}, there is no longer any need for L_{1s} to equal L_s. Now L_{1s} is no longer a part of the expression at all. With this restriction removed, it is possible to make $L_1 = L_s$ if, instead of the shield, we use a second conductor identical with lead 1. So that we may use as many previously derived formulas as possible, we refer to the second conductor as lead s. Then, to make $L_{s2} = L_{12}$, we twist leads 1 and s together so that an equal amount of flux from lead 2 links each twisted lead. The circuit appears in Figure 4-22.

A twisted pair is advantageous only with a balanced circuit. If we connect lead s like a shield, with both ends grounded, there is much less shielding effect unless we can make L_{1s} approximately equal to L_s. This is nearly impossible unless the shield fully encloses lead 1. If the circuit cannot be balanced, a true shield, enclosing lead 1, should be used.

Equality of R_1 and R_s depends on the tolerance of the respective circuit elements. Equality of L_1 and L_s depends on the physical placement of the leads and how tightly they are twisted. A perfect balance can never be achieved, but should be the goal. To achieve a 40-dB noise reduction due to the balanced circuit, I_c must be 40 dB less than I_d, or equal to $I_d/100$. This is possible only if R_1 (or $R_{1S} + R_{1L}$) differs from R_s (or $R_{sS} + R_{sL}$) by no more than 1 percent, which requires a tolerance of 0.5 percent on each impedance, R_{1S}, R_{sS}, R_{1L}, and R_{sL}.

As with a shielded lead, we may use a balanced circuit in lead 2 instead of lead 1, with equal effectiveness. Then we make R_s and L_s equal to R_2 and L_2, respectively, instead of to R_1 and L_1. Since leads s and 2 are twisted, $L_{1s} = L_{12}$. The current I_2 is not zero but is equal to I_s, so equal noise voltages appear across R_2 and R_s. This is known as common-mode noise. If we use a differential amplifier to sense the voltage, then it will ignore the common-mode noise, as desired. A balanced circuit is normally intended to reduce noise coupled to *and* from the circuit. Therefore, nearly every balanced circuit uses a differential amplifier to receive the signal. The differential amplifier must be insensitive to these common-mode voltages. The degree to which it ignores them is called the *common-mode rejection ratio*. Ratios of 60–80 dB can be achieved and are used in telephony. Any differences between R_s and R_2, or between L_s and L_2, will convert part of the common-mode voltage to a differential voltage across the amplifier inputs. This will degrade the common-mode rejection ratio.

We have introduced the balanced circuit as a method of reducing inductive coupling. Due to the analogy between electric and magnetic fields, discussed in Chapter 3, the analysis can easily be extended to capacitive coupling. If lead 1 is twisted with lead s, equal and opposite charges will appear along the leads. Because they are twisted, each lead has an equal capacitance to lead 2. Thus there will be no

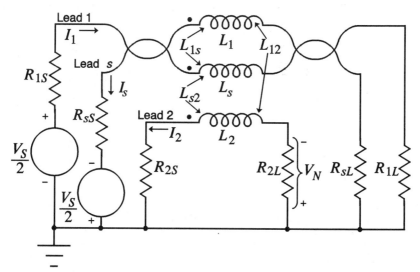

Figure 4-22   Balanced circuit

net charge induced onto lead 2, and therefore no capacitively coupled noise. If, instead, lead 2 is twisted with lead s, equal charges will be induced onto leads 2 and s. Therefore, equal voltages will appear on both leads, and the differential amplifier will detect no difference.

The most effective method of balancing a low-frequency system is to use two transformers, as shown in Figure 4-23. This also opens the undesired ground connection and blocks the conductive path for common-mode current. Such current does, however, flow through inter-winding capacitances of the transformers. This capacitive reactance is the common-mode impedance, which thus decreases with frequency. Therefore, the common-mode current increases with frequency, causing more harm at higher frequencies. In cables that intentionally carry radio frequencies, this inter-winding capacitance renders the scheme nearly useless.

Traditionally, the circuit shown in Figure 4-23 has been used in telephony. Although telephone lines do not intentionally carry radio frequencies, they can pick up stray radio signals which can affect the telephone equipment if not suppressed. The transformers provide an acceptable degree of isolation, except in the most severe cases. However, the transformer is now sometimes replaced by a transformerless *Subscriber Line Interface Circuit (SLIC)*. Such a circuit must be designed very carefully to achieve the same degree of balance. At radio frequencies it is very difficult to achieve the necessary balance, and other suppression methods, such as common-mode chokes, must be used.

4-8 EXPERIMENTAL MEASUREMENTS

The proof of the validity of the above formulas is their correlation with experimental measurements. Effectiveness of various shield types has been measured in experi-

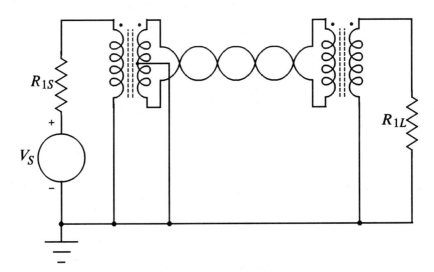

Figure 4-23   Transformer-coupled balanced circuit

ments by Ott[5]. His results confirm our earlier examples, and we now describe his experiments.

A difficult problem in experiments of this type is to guarantee that the ground return lead is located exactly as assumed in the calculations. The best approach is to eliminate the ground return lead entirely by collocating the source and load resistances R_{1S}, R_{1L}, R_{2S}, and R_{2L}. If we then arrange the cable in a circle, we may calculate the inductances using known formulas. To increase the coupled noise to an easily measurable level, the circle may contain several turns.

In Ott's experiments, lead 1 is a ten-turn coil of #20 wire, with a 9-in. diameter. The shield is on lead 2, which is a three-turn coil of shielded cable with a 7-in. diameter. Lead 2 is located in the same plane as lead 1, and concentric with it. Resistance $R_{2S} = 100\ \Omega$, $R_{2L} = 100\ k\Omega$, and I_1 is a 0.6-ampere current from a 50-kHz sine wave source. The frequency of 50 kHz is lower than most RF applications but convenient for these experiments. The setup is shown in Figure 4-24.

The self and mutual inductances for this geometry are defined as for any other geometry, but their calculation involves elliptic integrals unless some approximations are made. We will assume that the cross section of each coil is approximately circular and that its cross-sectional area A_x is proportional to the number of turns N. Then, since this area is proportional to the square of the cross-sectional diameter D_x,

$$D_x \approx d\sqrt{N} \qquad (4\text{-}47)$$

where d is the diameter of the wire or shield.

We now calculate L_1 and L_s using the following formula for a multi-turn circular coil whose cross section is approximately circular[6]:

$$L = \frac{\mu N^2 D}{2}\left(\ln \frac{8D}{D_x} - 2\right) = \frac{\mu N^2 D}{2}\left(\ln \frac{8D}{d\sqrt{N}} - 2\right) \qquad (4\text{-}48)$$

where D is the coil diameter. For loop 1, $N = 10$, $D = 9$ in., and $d = 0.032$ in. for #20 wire, so

$$L_1 = \frac{1.257\ \mu H/m \cdot 10^2 \cdot 9\ \text{in}}{2 \cdot 39.375\ \text{in/m}}\left(\ln \frac{8 \cdot 9}{0.032\sqrt{10}} - 2\right) = 65.6\ \mu H \qquad (4\text{-}49)$$

If lead 2 is RG/58U coaxial cable, then its center conductor is #20 wire and its shield diameter is approximately 0.113 in. For the shield loop, $N = 3$ and $D = 7$ in., so

$$L_s = \frac{1.257\ \mu H/m \cdot 3^2 \cdot 7\ \text{in}}{2 \cdot 39.375\ \text{in/m}}\left(\ln \frac{8 \cdot 7}{0.113\sqrt{3}} - 2\right) = 3.6\ \mu H \qquad (4\text{-}50)$$

The turns of the center conductor of coil 2 are not closely spaced but are separated by the shield diameter. Therefore, we cannot use the approximations mentioned above to calculate L_2 directly. Instead, we calculate the inductance L_d of the length

5. Ibid.

6. Ramo, Whinnery, and Van Duzer, *Fields and Waves,* p. 191.

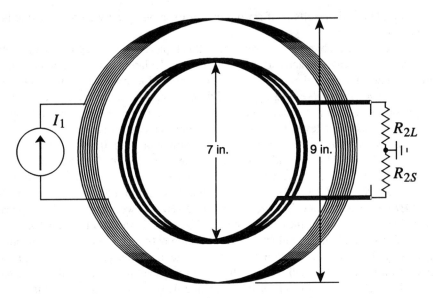

Figure 4-24   Experiment for which calculations verified results

of shielded cable, and note that it must be equal to $L_s + L_2 - 2L_{s2}$, which is the inductance of the differential-mode loop. Winding it into the 7-in. coil has negligible effect on L_d. We also recall that neglecting the inductance of the pigtail leads, $L_{s2} = L_s$, therefore, $L_2 = L_d + L_s$.

The length of the shielded cable is $3\pi \cdot 7$ in., or 65.9 in., and the inductance of RG58/U coaxial cable is 0.036 μH/ft. Therefore its total inductance L_d is approximately 0.2 μH, and

$$L_2 = L_s + L_d = 3.7 \ \mu\text{H} + 0.2 \ \mu\text{H} = 3.9 \ \mu\text{H} \tag{4-51}$$

As noted above, $L_{s2} = L_s$, and we also recall that $L_{12} = L_{1s}$. Therefore, the only inductance that we still must calculate is either L_{1s} or L_{12}. We will calculate L_{1s}, since here again we can use approximations and avoid evaluating elliptic integrals.

The mutual inductance between two concentric single multi-turn coils in the same plane[7] is

$$L = N_1 N_2 \mu \frac{D_1 + D_2}{2} \left[\left(1 - \frac{k^2}{2} \right) K(k) - E(k) \right] \tag{4-52}$$

$$\text{where} \quad k = 2 \frac{\sqrt{D_1 D_2}}{D_1 + D_2}$$

and where $E(k)$ and $K(k)$ are the two kinds of complete elliptic functions. Since $D_1 = 9$ in. and $D_2 = 7$ in., $k = 0.9922$, and, for a value of k this close to 1, we may use the following approximations:

---

7. Ibid, p. 190.

$$E(k) \approx 1$$

$$K(k) \approx \ln\left(\frac{4}{\sqrt{1-k^2}}\right) \tag{4-53}$$

Thus,

$$L_{1s} = 10 \cdot 3 \cdot \frac{1.257\ \mu H/m}{39.375\ in/m} \cdot \frac{9\ in + 7\ in}{2} \cdot \left[\left(1 - \frac{0.9922^2}{2}\right)\ln\left(\frac{4}{\sqrt{1-0.9922^2}}\right) - 1\right]$$

$$= 5.8\ \mu H \tag{4-54}$$

Various cable connections were tested for shielding effectiveness, and the results appear in the referenced work[8]. An ungrounded shield has no effect, as discussed earlier. To maintain the same physical spacing in all parts of the experiment, a shielded cable with the shield ungrounded is used to obtain the 0-dB reference level. All other measurements are given with respect to this level. Usually, adding a shield grounded at one end has no effect, so the coupling must be principally inductive. Tests on a shielded cable grounded at both ends yielded results similar to the examples in Section 4-4.

The most important result was an improvement of 80 dB, or a factor of 10^4, when one end of the *entire circuit* (not just the shield) was ungrounded. This forces all current to return via the shield and should eliminate the noise. The noise that does remain is due to the capacitance of the ungrounded end of the shield, which is unavoidable. So the best approach is a simple coaxial cable with the shield serving as the *only* ground return for the circuitry at one end.

Another noteworthy point is that there is little difference between coaxial cable and shielded twisted pair *with one side grounded at both ends*. This might be expected, since all that is really changed is the cross section of the cable. Since coaxial cable usually has better transmission characteristics, it is usually more desirable than a shielded twisted pair *connected in this manner*.

Tests of an unshielded twisted pair, with one conductor connected to ground, confirm the results of Problem 15. Coaxial cable performs much better, even with both ends grounded. The reason, of course, is that we can no longer assume $L_{1s} = L_s$, and the analysis of Section 4-2 is not valid. The grounded conductor has little shielding effect. A *balanced* twisted pair does perform well but works on a different principle, with equal and opposite noise voltages in the two leads.

We must remember that shielding ineffectiveness is often due to the inductance of pigtail leads (see Problem 9). This is particularly true at high frequencies. Therefore, to correlate theoretical and experimental results, the inductance of the pigtail leads must be calculated. Except with simple geometries such as Figure 4-17, this is usually very difficult. It was not mentioned in Ott's experiments, which were conducted at a relatively low frequency. If pigtail leads are eliminated from high-frequency experiments, the results may be unrealistic. To include the effects of

---

8. Ott, *Noise Reduction Techniques,* pp. 56–59.

pigtail leads, one may correlate the calculated results using a simple geometry, such as a circle. By moving the pigtail leads, one may observe their effect on the coupling.

4-9 PROBLEMS

1. Write the loop or node equations for Figure 4-4 with $R_{2P} = \infty$ ($I_2 = 0$) and find the transfer function V_N/V_S. Show that for large L_p it reduces to Equation (3-4) with $R_{2P} = \infty$. At what frequency can V_N become greater than V_S? Why?

2. In Figure 4-4, let $C_{1s} = 100$ pF, $C_s = 31.6$ pF, $C_G = 4.8$ pF, $L_p = 2$ μH, $R_{1S} \approx R_{1P} = 100$ Ω, and $R_{2P} = \infty$. What is the magnitude of V_N/V_S at 7 MHz?

3. Draw a diagram like the one in Figure 4-4 for a circuit in which the "victim" lead (lead 2) is shielded, instead of the source lead. Compute the transfer function by writing the loop or node equations. Why must it be identical with the transfer function obtained in Problem 1?

4. Derive Equations (4-1) and (4-2).

5. Write the loop equations for the circuit of Figure 4-7 and show that it is equivalent to that of Figure 4-6.

6. Redraw Figure 4-10 and include the vertical scale in decibels.

7. Plot Equation (4-15) on log/log axes, showing vertical and horizontal scales.

8. Solve Equations (4-3) through (4-5) for I_2, assuming $L_{12} = L_{s2}$ but now with $L_s > L_{1s}$ and with finite R_1 and R_2. Show that the expression for I_2 is identical to Equation (4-11) except that Z_s replaces R_s.

9. In Figure 4-17, the termination leads are shortened so that $r_p = 1$ cm. Calculate the shielding effectiveness, assuming $R_s = 0$.

10. In Problem 9, what percentage of the center conductor current will return via the common ground instead of the shield?

11. In Figure 4-17, with the dimensions given in the text, a 50-μH common-mode choke is installed. Calculate the shielding effectiveness, given the following:
   (a) the coupling coefficient of the choke is 0.95;
   (b) the coupling coefficient of the choke is 1.

12. An AC source is connected to a resistive load via a shielded cable grounded at both ends, as shown in Figure 4-25. All dimensions appear in the figure, and the segments are labeled a through k for identification.
   (a) Calculate the approximate total inductance, L_p, of the two pigtail loops a-b-c-d and f-g-h-j. Assume no mutual inductance between them (the total inductance is equal to the sum of the two individual inductances). Neglect the contribution due to the small segments b, d, f, and h. Also neglect field "fringing" at the tops and bottoms of these loops; assume that the flux defining L_p does not vary in the vertical direction.
   (b) Similarly, calculate the approximate inductance, L_s, of the shield loop c-e-j-k. Neglect the contribution due to segments c and j, and assume that the flux defining L_s does not vary in the horizontal direction (neglect fringing).

13. In Problem 12, assume that the flux due to the current in segments a and g links only loop 1, and that the flux due to the current in segments c and j links only the shield loop.

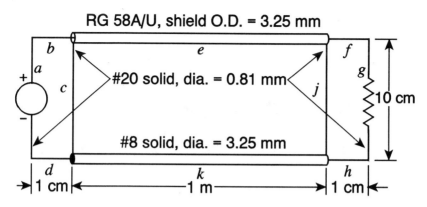

Figure 4-25   Problem 4-12

Using the calculated values of L_p and L_s, and the method from Section 4-4, calculate L_{1s}, the mutual inductance between loop 1 and the shield loop.

14. In Problem 12, the source is 1 V at 10 MHz, and the load is a 50-ohm resistor. Calculate the magnitudes of the following:
    (a) the shield current, I_s;
    (b) the ground current, $I_1 - I_s$;
    (c) the net magnetic flux linking the shield loop. (Neglect the contribution from the pigtail leads themselves.)

15. Even after reading this chapter, a certain engineer is convinced that a twisted pair with one lead grounded is an effective shield. He twists a conductor around a signal lead located 5 in. above its common ground return, and grounds both ends. The separation between the twisted leads is 0.05 in. All conductors are #30 AWG, with a diameter of 0.01 in. Assume that the twisting does not affect L_s or L_{1s} and calculate their values using the formulas from Chapter 3. Calculate the "shield" transfer impedance and the shielding effectiveness, assuming R_s to be negligible.

5

GROUNDING OF
MULTIPLE-CHASSIS SYSTEMS

We noted in Chapter 4 that it is desirable to provide a separate ground return path for each signal lead. If a signal lead is shielded, then the shield should be the only ground return path for that lead. When the source and load are in the same rack or cabinet, this is only a minor inconvenience. The only added cost is that of the return wire and usually a differential amplifier or a transformer. When two or more signal leads interconnect two cabinets, however, the problem becomes more complex. If each signal lead has its own ground return, then there are multiple ground points in one cabinet. These ground points must not then be interconnected except by the external ground return leads; otherwise, the return currents do not flow as intended. We shall now study methods by which the return currents can be controlled.

5-1 SIGNAL GROUND CONNECTIONS

A circuit designer normally assumes that a perfect ground is available wherever he needs it. Since he is probably responsible for the design of only one cabinet, this equates to assuming a zero potential on every ground lead in the cabinet. The first step, then, is to assure this by using the methods of Chapter 4 within each cabinet. Ideally, there is a separate ground return lead for each high-frequency signal lead running entirely within the cabinet. The ground return may be a shield around the signal lead or, for a balanced circuit, a lead twisted with the signal lead. As shown in many figures in Chapter 4, the return lead should then be connected to the chassis or cabinet at only one point.

Signal currents running between cabinets can be analyzed using the methods of Chapter 4, with larger distances. Two equipment cabinets, with two signal leads running between them, are shown in Figure 5-1. We assume that both signal leads carry high frequencies and therefore must be shielded. Each signal current should preferably return via a path as close to that signal lead as possible. This minimizes the area of each signal loop and thus reduces the flux linking both loops, as desired.

If the loads are grounded to the chassis, as shown by the dotted connections, then V_{g1} and V_{g2} are zero, as the designer assumes. However, the return current for signal lead 1, for example, will then divide itself among all possible paths. Let I_{s1} represent the portion that returns via the shield, as intended. Usually $I_1 \neq I_{s1}$, so even if the shielded cable is perfect there will be some magnetic flux outside shield 1. If the shield diameter is much less than the cable separation distance, the flux density outside the shield is

$$\mathbf{B}_1 = \frac{\mu(I_1 - I_{s1})}{2\pi r}\mathbf{a}_\phi \qquad (5\text{-}1)$$

where r is the distance from the axis of lead 1 and $\mathbf{a}_\phi$ is a unit vector in the direction determined by the right-hand rule. Thus, the flux outside the shield is proportional to $I_1 - I_{s1}$. If the dotted connection shown on lead 2 is not present, then lead 2 and its shield form an isolated loop. If that shield is perfect, then the area of that loop is zero, and the flux induces no voltage into the loop. If the dotted connection is present on lead 2, there are multiple loops, which we must consider separately.

Let A_2 represent the area enclosed by lead 2 and the common ground return at the bottom of Figure 5-1. The shield and the common ground return also enclose the area A_2, since we assume the shield to be perfect. Therefore, the flux induces equal voltages into loop 2 and its shield loop, each voltage being

$$V_2 = V_{s2} = \int_{A_2} \mathbf{B}_1 \cdot d\mathbf{A} \qquad (5\text{-}2)$$

Since loop 2 and its shield do not form a balanced circuit, the self impedance of loop 2 is not equal to that of the shield loop. The two equal induced voltages then

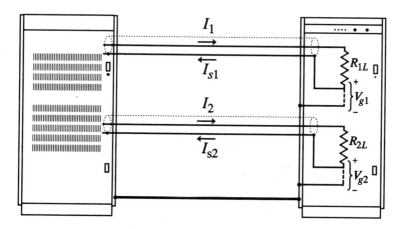

Figure 5-1   Signal and return leads running between cabinets

cause unequal currents to flow in loop 2 and its shield loop, and the currents in turn cause unequal voltage drops. The net noise voltage appearing across R_{2L} is the difference of these voltage drops, which is not zero. Thus, a noise voltage appears across R_{2L} due to $I_1 - I_{s1}$, although we assume both shields to be perfect. By similar reasoning, a noise voltage appears across R_{1L} due to $I_2 - I_{s2}$.

It is better to avoid grounding the loads, thus omitting the connections shown as dotted lines in Figure 5-1. With the loads not connected to ground, $I_{s1} = I_1$ and $I_{s2} = I_2$, so there is no magnetic flux outside the shields. Each loop is isolated and has zero area, so even if there were any flux, it would induce no noise voltage. Noise coupling is thus reduced for two reasons. However, V_{g1} and V_{g2} are then not normally zero. To detect the true signal voltage requires a transformer or differential amplifier. With either of these devices, load 1 responds only to the difference between its signal voltage and V_{g1}, and similarly for load 2. The voltages V_{g1} and V_{g2} do not then affect the received signal.

If a load must be grounded to the chassis, then we must use other means to force the signal current to return via the shield. We may add a common-mode choke to each signal lead and its respective ground return, as discussed in Chapter 4. This increases the impedances of the undesirable return paths, but not the desired return path. Thus, more current returns via the desired path. It also equalizes the self inductances of each lead and its respective shield, so that equal voltages induce equal currents. It thus causes the circuit to behave more like a balanced circuit. The common-mode chokes should be as large as possible without signal degradation due to leakage flux, and within cost limitations.

Systems have been designed in which an entire chassis is grounded only via the cable shields. This corresponds to keeping the dotted connections in Figure 5-1, but eliminating the solid connection between the cabinets at the bottom of the figure. The scheme is shown in Figure 5-2. It does force all signal current to return via the shields, since there is no other path. This reduces radiated emissions, which will be discussed in later chapters. However, some signal current may return via a shield other than its own. This current flows through parts of two signal loops, and the flux that it generates links both loops. This flux causes noise coupling between those loops. Therefore, this scheme is less effective in reducing cross-coupling between loops. It does have some effect, since the mutual inductance between loops is less when the ground connection is not present.

The scheme is sometimes difficult to implement, since any stray connections to the ungrounded chassis would divert some return current and destroy the integrity of the system. One cabinet must be grounded for safety, and then the high-frequency voltage V_g appears on the other cabinet. That cabinet must be insulated from any steel framework used in the building, and from all water pipes and ventilation ducts. Usually, someone must periodically check the system integrity by disconnecting all shields and checking for no continuity to ground. A stray ground fault can cause erratic operation that is very difficult to trace.

Even with no connection, some capacitance C_{cg} exists between the cabinet and the building framework. We may estimate its magnitude using a formula for a

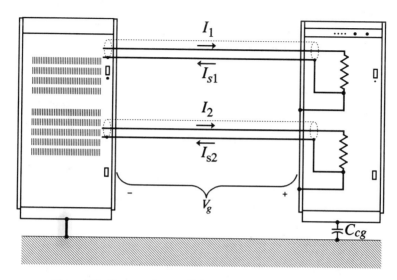

Figure 5-2 Return path controlled by isolating cabinet from ground

simpler geometry that approximates the actual dimensions. For example, if the cabinet is mounted on recessed casters that support it 1 cm above a metal floor, and if the cabinet occupies $1/3$ m^2 of floor area, then

$$C_{cg} = \frac{\epsilon_0 \cdot 1/3 \text{ m}^2}{0.01 \text{ m}} = 295 \text{ pF} \approx 300 \text{ pF} \qquad (5\text{-}3)$$

If we know the lead spacings and lengths, we may use this capacitance value to estimate the cross-coupling between two leads.

A variation of this scheme is to ground the cabinet but not connect any circuitry to it, as shown in Figure 5-3. Instead, we connect all grounds inside the cabinet to a common busbar and connect the cable shields to the busbar but not to the cabinet. The effect is the same as above, and the hardware is often easier to implement. The cable shields, however, must not be connected to the cabinet containing the isolated busbar, since the high-frequency voltage V_g exists between them. We will observe in Chapter 11 that isolating the cable shields from the cabinet prevents the cabinet from functioning as an effective shield, if needed. The capacitance of concern here is the busbar-to-cabinet capacitance C_{bc}, also shown in Figure 5-3. It is usually smaller than C_{cg} unless the building framework is non-metallic.

A problem can occur when a ready-built instrument is to be installed inside the cabinet. The metal instrument case is usually permanently connected to its signal ground, and the signal ground is the busbar. If we mount the instrument in the cabinet of Figure 5-3, a connection results between the internal busbar and the cabinet. The instrument case then causes an undesired alternate path for the signal return currents. The instrument may be completely unrelated to the external cables,

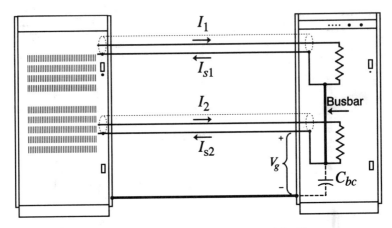

Figure 5-3 Return path isolated from grounded cabinet

and may not even be located near them. To block the undesired path, we must insulate the instrument case from the cabinet in which it is mounted. If the instrument case is exposed, a temporary short circuit between it and the cabinet is likely. Such a short may result simply from touching both surfaces with a metallic tool such as a screwdriver or pliers. This may cause intermittent problems that are very difficult to locate.

An instrument designed for cabinet mounting may contain a removable connection between its signal ground and its case. With the scheme of Figure 5-3, this connection should be removed when the instrument is permanently mounted in the larger cabinet. Then the instrument case need not and should not be insulated from the cabinet. To permit this type of isolation, rack-mounted instruments should be designed with this removable connection.

5-2 SAFETY GROUND CONNECTIONS

Most cabinets contain equipment powered by the AC line or another high-power source. If such a cabinet is metal, there is always the possibility of an internal short circuit to the cabinet. Such a short circuit may have disastrous consequences unless the cabinet is grounded by a conductor heavy enough to carry the fault current without melting. Even in normal operation, both sides of the AC line are usually connected through small capacitors to the metal cabinet, to bypass common-mode noise on the line. This allows some 60-Hz current to flow from the ungrounded side of the AC line onto the metal cabinet. Unless the cabinet is grounded, an AC voltage as high as 60 V (half the line voltage) will appear on the cabinet. This is disturbing to the user, and depending on the capacitor size, may also be hazardous.

In the scheme shown in Figure 5-2, the cable shields provide the only ground on the "ungrounded" cabinet. Their pigtail connections may not be heavy enough to carry a 20- or 30-ampere fault current. Furthermore, if the cables become disconnected for any reason while the AC power is still connected, the cabinet loses its ground. A hazardous condition then is created by the current through the line bypass capacitors or a possible short circuit. To prevent this danger, the National Electrical Code requires that every metal cabinet containing equipment powered from the AC line must be grounded. The ground conductor must be large enough to carry the current for which the AC equipment is fused. This usually requires a #12 or larger wire for a cabinet of typical size. If we add such a wire to Figure 5-2, it provides another return path for the signal current, which is what we are trying to prevent. The safety grounding requirement thus conflicts with the desirability of not grounding the cabinet.

One solution to this problem is to use the scheme shown in Figure 5-3 instead. The safety ground is then connected only to the cabinet. Since the busbar is not exposed, it does not require a safety ground.

If the connection between the two grounds is unavoidable, as in Figure 5-2, then we may insert an RF choke in series with the safety ground. The choke then impedes the signal currents from following this undesired path. Its value is effectively added to L_1, L_s, and L_{1s}, so it behaves in the same way as a common-mode choke. It must, however, be able to carry the fault current for which the circuit is fused, to satisfy code requirements. A choke wound with wire of this size is bulky, even for small inductances.

5-3 PROBLEMS

1. Coupled noise is being detected in a two-cabinet system similar to the one shown in Figure 5-2. To analyze the problem, an oscilloscope must be connected to the circuitry in the ungrounded cabinet. The only available oscilloscope is AC powered and has a three-prong grounding plug, which, for safety reasons, must not be disconnected. Whenever the oscilloscope is connected, the problem disappears and cannot be analyzed. Explain.

2. A certain installation is similar to the one shown in Figure 5-2 but contains only one cable. The loops formed by the shielded cable and the ground return path through C_{cg} are labeled as in Chapter 4. Their inductances are $L_1 = 10 \ \mu H$, $L_s = 9.9 \ \mu H$, $L_{1s} = 9.8 \ \mu H$ ($L_{1s} < L_s$ due to leakage inductance of pigtail leads), and $C_{cg} = 300$ pF. The cable is driven by a 1-MHz signal.
   (a) Draw the equivalent circuit.
   (b) Calculate the percentage of return current diverted through the ground.

3. In Problem 2, a safety ground is added to comply with the electrical code. A 30-μH choke, rated for the fault current, is placed in series with the safety ground.
   (a) Add this choke to the equivalent circuit of Problem 2.
   (b) Calculate the percentage of return current now diverted through both grounds.

4. To save money, an engineer proposes a single toroid around all shielded leads, instead of a common-mode choke on each lead. The size of the toroid is such that it imparts the same inductance to each lead as would the individual chokes. What is the major disadvantage of this method?

6

LAYOUT AND GROUNDING OF PRINTED WIRING CARDS

Under some conditions, the material in Chapters 3 and 4 becomes applicable to systems as small as a single printed wiring card. If the frequencies are high enough, a current loop on a printed wiring card (*PWC*) can contain enough inductance to cause cross-coupling between loops. Capacitive coupling also may be a problem, although it is easier to correct. The analyses in Chapters 3 and 4 assumed nearly equal voltages and currents in the source and victim loops. This was a valid assumption for interference between similar circuits.

The most likely cause of interference on the same PWC, however, is a high current inducing a voltage into a nearby low-voltage circuit. This is common when two dissimilar circuits are close together. It immediately suggests that a circuit carrying a large current should not be located near a circuit that is sensitive to low voltages. The first step, then, is to classify circuits by voltage and current magnitudes and to separate physically the different classes of circuits.

6-1 TYPES OF CIRCUITS

Nearly every electric circuit and its components can be classified into one of five types. These are audio, radio-frequency (RF), digital, power conversion, and electromechanical. Each has different typical voltage and current levels and different frequencies.

Audio circuits operate over a frequency range from DC to approximately 20 kHz. Voltage levels may be in the millivolt range, so the circuits are typically very sensitive to externally induced voltages. Except in final stages of power amplifiers, the voltages and currents are usually not large enough to cause interference to other circuits. The chief concern with audio circuits is their susceptibility to nearby currents and sometimes to voltages.

Radio-frequency circuits operate above 10 kHz, with virtually no upper frequency limit. Voltage and current ranges are comparable to audio circuits. Like audio circuits, low-power RF circuits are often sensitive to externally induced voltages. High-power RF circuits, on the other hand, generate fields that may interfere with other circuits. Thus, with RF circuits, both emissions and susceptibility are of concern.

Power conversion circuits often operate at the 60-Hz line frequency. Some, however, use high-frequency switching to allow smaller transformers. In either case, currents are usually large. With high-frequency switching, rise times are usually made as short as possible to avoid wasting power. The result is a high current being very rapidly switched, which is very likely to cause interference to other circuits. In this respect, power conversion circuits exhibit the same interference problems as other high-power RF circuits.

Digital logic circuits operate over a wide range of frequencies. Their voltage levels are higher than most audio and low-power RF circuits, so interference from those circuits is not a major concern. Interference to digital circuits is usually from power conversion or other high-power RF circuits. If many digital logic gates are repeatedly switched in synchronism, the resulting field may interfere with other circuits. With digital circuits, this occurrence depends on the data being processed and may be sporadic.

Electromechanical circuits include relays and mechanical switch contacts. Voltages and currents may be very large compared to those in the other circuit types discussed above. Although we might expect that the frequencies involved must be very low, this is only true of the steady-state repetition rate. An opening electrical contact causes a very sudden interruption of current. The circuit must contain some inductance, and if it is appreciable, the contact may spark. A sparking contact generates voltages and currents that may easily interfere with other circuit types. Sparking can be suppressed with a capacitor, varistor, or diode. However, if such a device should fail, an interference source may develop that is very difficult to locate. Electromechanical contacts are primary causes of interference to other circuits but are not themselves susceptible.

An exception to the general definition of electromechanical circuits is a contact that remains in a fixed state during normal operation. A fuse, for example, is an electromechanical contact. When a fuse opens because of an overload or short circuit, large moving magnetic fields result, and a great danger of coupled noise exists. However, since a blowing fuse is regarded as an abnormal occurrence, we do not consider the fuse to be an electromechanical device and we may collocate it with any circuit type. If a circuit is to be fail-safe, however, then another portion of the circuit must continue to function even while a fuse is blowing. In such a circuit, a fuse must be considered as we would consider any other electromechanical device. A similar discussion applies to *DIP* (*D*ual *I*n-line *P*ackage) switches, which are often used to set parameters before energizing a circuit. If they are not normally switched during circuit operation, we do not regard them as electromechanical devices.

6-2 LAYOUT CONSIDERATIONS

Often, a PWC contains only one of the above circuit types. If it contains more than one type, the types should be grouped separately. Preferably, the layout engineer should allow a space between circuits of different types, but often this is not possible. Such cases may require a shield between different circuit types, as discussed in later chapters.

All leads running within a circuit, or between circuits of the same type, should be kept away from other circuit types and their associated leads. This may imply longer leads, but is still preferable. A suitable layout of a typical system is shown in Figure 6-1. All circuitry of each type is arranged into a rectangular area. The shortest lead connecting any two components within a rectangle must lie completely within the rectangle. Thus, laying out the circuits in rectangular blocks will inherently keep their internal leads away from foreign circuits. If any circuits must be arranged in blocks containing concave boundaries, intra-circuit leads should be kept within the boundaries even if the route is not direct.

Of course, the circuit types must usually be interconnected. Such interconnections should be minimized and may require buffers or other special treatment. The functions of the various circuits imply what interconnections are necessary between

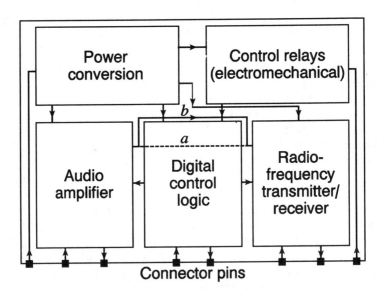

Figure 6-1 Dissimilar circuit types interconnected on same PWC

them. For example, the power source for the PWC must be connected to the power conversion circuitry, which then supplies the required voltages to all other circuits. Similarly, the digital logic controls the audio and RF circuitry, but is itself controlled by relays. The interconnections required for the circuit shown in Figure 6-1 are visible in the figure. Each line may represent several leads routed via the same path.

The circuit blocks are arranged so that most leads running between different circuit types need not pass near a third type. There is one group of leads, however, that must interconnect the audio amplifier and the transmitter/receiver, which are not adjacent. The figure shows two routing alternatives labeled *a* and *b*. The optimum route for an interconnecting lead is the route that produces the least electric and magnetic flux in all circuits of dissimilar types. A current in lead *b* would obviously produce less flux near the digital logic than would the same current in lead *a*. The digital logic thus has a lower mutual inductance with respect to lead *b* than lead *a*. Although longer, lead *b* picks up less noise. This, then, is the preferred route for all leads that must interconnect the audio amplifier and the transmitter/receiver in Figure 6-1. The shortest route is not necessarily the best, unless it also injects the least flux into any foreign circuits. This might occur if a lead passes near a foreign circuit for only a short distance.

6-3 CURRENT RETURN PATH

We have observed that coupling between circuits is minimized when current returns via a path as near as possible to the incident path. In Chapter 4, we tried to achieve this by forcing the current in a cable to return via its shield. On a PWC, the printed leads usually are not shielded. However, we can still minimize coupling by controlling the current return path. Ideally, the current should return by an identical path on the opposite side of the PWC. If an actual lead is placed on that side of the PWC, and if all incident current must return through this lead, then the arrangement is known as a *stripline*. A small section of a stripline is shown in Figure 6-2.

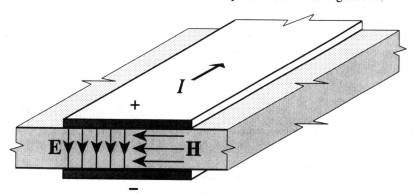

Figure 6-2 Stripline

If the PWC dielectric is much thinner than the width of each conductor, then the fields between the conductors are uniform except near the conductor edges. Let V_1 be the potential difference between two conductors of width w separated by a dielectric of thickness t. Let I_1 be the current flowing in them. In the dielectric between the conductors, except near their edges, the field magnitudes are

$$|\mathbf{E}_1| = V_1/t \qquad \text{(6-1)}$$

and

$$|\mathbf{H}_1| = I_1/w \qquad \text{(6-2)}$$

For example, in a stripline 0.1 in. wide on a dielectric slab 0.01 in. thick, the magnetic field between the conductors due to a 1-ampere current is

39.4 inches = 1 meter

$$|\mathbf{H}_1| = \frac{1\ \text{A}}{0.1\ \text{in}} \cdot 39.4\ \text{in/m} = 394\ \text{A/m} \qquad \text{(6-3)}$$

For a nonferrous dielectric, $\mu = \mu_0 = 4\pi \cdot 10^{-7}$ H/m, and therefore

$$|\mathbf{B}_1| = 1.257\ \mu\text{H/m} \cdot 394\ \text{A/m} = 495\ \mu\text{T} \qquad \text{(6-4)}$$

where 1 Tesla (T) = 1 Wb/m^2. The total flux between the conductors, per meter of stripline length l, is

$$\frac{\Phi_1}{l} = |\mathbf{B}_1|t = \frac{495\ \mu\text{T} \cdot 0.01\ \text{in}}{39.4\ \text{in/m}} = 0.125\ \mu\text{Wb/m} \qquad \text{(6-5)}$$

The fields outside this region are much weaker. Since all magnetic flux must follow a closed path, there must be as much total magnetic flux outside the conductors as between them. However, since the outside flux is spread over a large region, its density is much less than between the conductors. An exact solution is possible using the methods of Chapter 3, but we will not pursue it here. Instead we estimate the field intensity as follows. We enclose the conductor edges with a small imaginary cylindrical surface whose axis is midway between the edges, as shown in Figure 6-3. Its diameter must be greater than the dielectric thickness, so that both conductor edges are inside the cylinder. If the cylinder radius is less than the conductor width, as shown in the figure, then the surface passes through the conductors, and all of the magnetic flux Φ_1 must pass through this surface. We will estimate $\mathbf{B}$ by assuming it to be constant and normal to the surface everywhere. Then the average flux density is equal to the total flux Φ_1 entering the cylinder, divided by the area a of its surface. If the cylinder radius were greater than the conductor width, the cylinder would completely enclose the conductors. Then some flux would be completely inside the surface, so the average flux density passing through the surface would be less. The average flux density cannot exceed Φ_1/a in either case. At a distance $d = 1/4$ in. from the edge of the stripline, for example, the average magnetic flux density, designated $\mathbf{B}_{21}$, is no greater than

$$|B_{21}| = \frac{\Phi_1}{a} = \frac{\Phi_1/l}{2\pi d}$$

$$= \frac{0.125 \ \mu Wb/m}{6.28 \cdot 0.25 \ in} \cdot 39.4 \ in/m$$

$$= 3.136 \ \mu T \qquad\qquad\qquad \textbf{(6-6)}$$

If a second identical stripline is at this position, the magnetic flux linking it is equal to the flux density times the area of that loop, which is

$$\Phi_{21} = |B_{21}|tl \qquad\qquad\qquad \textbf{(6-7)}$$

By Faraday's law, the voltage induced into the second stripline is then equal to

$$|V_2| = \omega \Phi_{21} = \omega|B_{21}|tl \qquad\qquad cross-section \qquad \textbf{(6-8)}$$

For the above example, if the length of the stripline is 4 in. and the operating frequency is 1 MHz, the induced voltage is

$$|V| = \omega|B|tl$$

$$= 6.28 \cdot 1 \ MHz \cdot 3.136 \ \mu T \cdot \frac{0.01 \ in \cdot 4 \ in}{(39.4 \ in/m)^2}$$

$$= 0.0005 \ V \qquad\qquad\qquad \textbf{(6-9)}$$

This example dramatically illustrates the effectiveness of a stripline. Whenever the conductor width w is much greater than the dielectric thickness t, the coupled noise

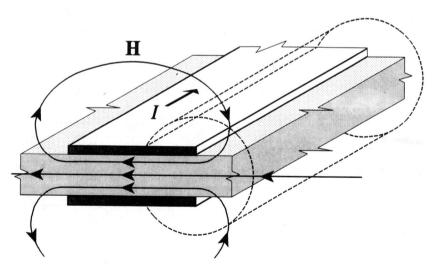

Figure 6-3   Imaginary surface enclosing edge of stripline

$$E_{RMS} = 4.44 \ \Phi N_2 f$$

is proportional to their ratio t/w if all other parameters are held constant. For any two-dimensional problem such as this, the electric field behaves the same as the magnetic field. A stripline is the ideal method of reducing coupled noise, since it minimizes the stray electric and magnetic fields.

Unfortunately, it is often difficult or impossible to force the current to return via the desired path. Unless the component connected to the lead has only two terminals, the current may split between two leads or combine with another current. Then the returning current will be unequal to the incident current, and the stray fields will not completely cancel. This greatly reduces the effectiveness of the stripline.

We may emulate the effect of a stripline by providing a solid conductive *ground plane* on the entire underside of the PWC. Assume, for the moment, that there are no holes in the PWC or the ground plane. This, of course, would require surface-mounted components. A cross-sectional view of this scheme appears in Figure 6-4.

The incident current I is flowing from left to right, and it returns via the ground plane. If the conductivity of the ground plane is perfect, or nearly so, the plane forms an equipotential surface for both the electric scalar potential V and the magnetic vector potential **A**. No matter where the incident current flows, the return current must distribute itself such that V and **A** are constant over the ground plane. The resulting electric and magnetic potential distributions are just as though there were an image of the incident current returning in a conductor or component below the ground plane. Figure 6-4 shows the electric field **E** between the conductors and the ground plane. The magnetic field in this region, although not shown, is directed into the page. For clarity, the figure is not drawn to scale, but we assume that the PWC thickness is much less than the conductor width. Then, nearly all of the electric and magnetic field is located in the region between the conductors and the ground plane. Outside this region, both fields are much weaker and are unlikely to cause noise coupling. This is theoretically an easy method of forcing the current to return via the desired path. It is very effective in reducing coupled noise.

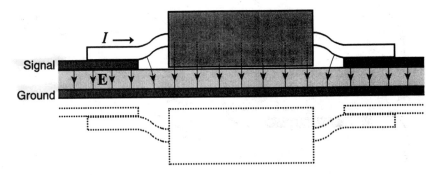

Figure 6-4 Fields near a surface-mounted component

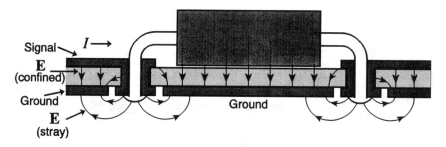

Signal

$I \rightarrow$

E
(confined)

Ground

E
(stray)

Ground

Figure 6-5   Fields near a component mounted through holes in PWC

Surface-mounted components, however, are sometimes inconvenient to use. Conventionally mounted components require mounting holes in the PWC and therefore in the ground plane, as shown in Figure 6-5. Near these holes, the stray electric field becomes significant and the ground plane has less effect. For the voltages encountered in most solid-state circuits, however, the electric field is insufficient to cause noise coupling. If each hole in the ground plane contains only a single lead, no current can flow inside the hole, so there is no stray magnetic field. Thus, a single component lead penetrating the ground plane usually causes no noise coupling, except with high-voltage circuits.

A much greater concern exists when there is a slot in the ground plane and a signal conductor is located in the slot, on the "wrong" side of the PWC, as shown in Figures 6-6 and 6-7. This is sometimes done to provide a connection across an intervening conductor, and it is called a *stitch*. Current then flows where there is no ground plane under it. The return path must then be much farther from the incident current than it would be with a stripline. Stronger electric and magnetic fields exist near the stitch, as shown in Figure 6-7. The danger of coupled noise is much greater.

Since the conductor separation is now large compared to their width, we may approximate the conductors as two thin wires. For the same conductors now sepa-

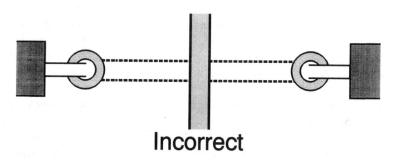

Incorrect

Figure 6-6   A stitch in a PWC (undesirable)

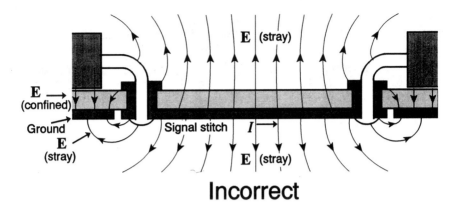

Incorrect

Figure 6-7   Fields near a stitch (undesirable) in a ground plane

rated by a distance $s = 0.2$ in., for example, the field at a distance of $1/4$ in. from the nearest conductor, in the same plane as the conductors, is approximately

$$|\mathbf{B}_{21}| = \frac{\mu I_1}{2\pi} \left(\frac{1}{d} - \frac{1}{d+s} \right)$$

$$= \frac{1.257 \; \mu\text{H/m} \cdot 1 \; \text{A}}{2\pi} \cdot \left(\frac{1}{0.25 \; \text{in}} - \frac{1}{0.45 \; \text{in}} \right) \cdot 39.4 \; \text{in/m}$$

$$= 14 \; \mu\text{T} \tag{6-10}$$

Comparing Equation (6-10) with Equation (6-6) shows that for the dimensions in this example, the stitch causes the stray magnetic field to increase by a factor of five. Since Equation (6-6) gives only an *upper limit* for the flux with no stitch, the actual degradation may be even greater.

The effectiveness of a ground plane is reduced in this manner whenever the return current is not a mirror image of the incident current. The ground plane must completely cover one side of the PWC, allowing holes only for individual component leads. Wherever a conductor crosses a slot in the ground plane, the return current must deviate, and the magnetic field increases as shown in the example of Equation (6-10). This point is often difficult for designers and layout specialists to understand. When a stitch is necessary, it is always because of an intersecting lead (not shown in Figure 6-7), and this intersecting lead then crosses the resulting slot in the ground plane. Both leads then pass over an area where the return currents must deviate. On a PWC with a ground plane, it is usually much better to reroute the intersecting lead, even if this requires a very roundabout path. A correct connection, with no stitch, appears in Figures 6-8 and 6-9. The reduction in the stray electric field is evident in Figure 6-9 compared to Figure 6-7. The magnetic fields are perpendicular to the page and therefore are not visible, but the reduction is just as significant.

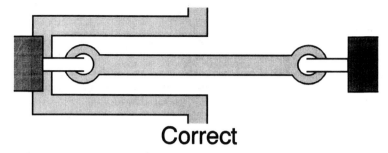

Correct

Figure 6-8   PWC conductor rerouted to avoid a stitch

If a stitch is absolutely necessary, a multilayered PWC will alleviate this problem. When two leads must cross, the stitch may then be made using a layer other than the ground plane. If the ground plane remains intact, the return currents can follow the mirror images of the incident currents, and the stray magnetic fields are minimized. With multilayered cards, however, the designer must avoid running a lead directly over another. If he does, the capacitance and mutual inductance between the leads become so large that the stripline is less effective. Two leads in different layers should cross at right angles. If they run in parallel paths, they should be separated just as though they were in the same layer. A suitable approach is to place all leads running in one direction on the same layer, and all leads perpendicular to them on another. This is sometimes known as xy routing. If the ground plane is located between these two signal layers, there can be no coupling where the leads cross, and the coupling is minimized.

For high-volume applications where the cost of a ground plane is significant, a ground grid may be used instead. The best type of grid includes a ground conductor under every signal lead. The return current then can flow under the incident current, and the effect is comparable to a solid ground plane. It is not exactly equivalent, however, because in a solid ground plane the return current spreads out under each signal conductor. This is impossible with a grid.

Another type of ground grid uses a fixed pattern of squares or rectangles. Unless all signal leads pass over the grid conductors, the return currents cannot

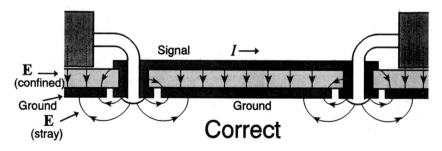

Correct

Figure 6-9   Stray fields reduced by avoiding a stitch

always flow directly under the incident currents. This type of ground grid is therefore usually less effective.

When two or more circuit types share the same PWC, as discussed in Section 6-2, each circuit type should have its own ground plane. Here it is undesirable to allow the return current from one circuit type to flow in the ground plane of another. This can reliably be prevented only by separating the ground planes. No ground plane should extend into the area of another circuit type.

Where two circuit types are interconnected, however, a return path is necessary for the current flowing between the circuits. Therefore, some type of connection is necessary between the ground planes. The return current must be kept equal to the incident current, which may be difficult to achieve. As discussed in Chapter 4, intentional common-mode impedance often helps keep these two currents equal. Examples include transformers, optical isolators, or common-mode chokes. Each interconnection normally requires individual consideration. This is another reason to keep such interconnections to a minimum.

6-4 POWER DISTRIBUTION WITHIN A PRINTED WIRING CARD

A circuit designer usually assumes that power supply voltages, like circuit grounds, are available wherever needed. Again, this is not true and the actual power distribution path must be considered. Practical power distribution systems contain series impedance and shunt admittance. They can pick up noise just as signal paths do.

The load current drawn from a power bus usually varies, often at a high frequency. The varying load current behaves like a current source connected at the load end of the bus, which is typically the power pin of an integrated circuit (IC). A schematic representation of a power distribution bus appears in Figure 6-10, where I_L represents the DC component of the load current and I_T represents its transients.

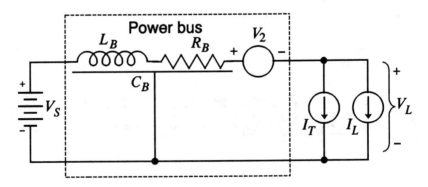

Figure 6-10 Equivalent circuit of a power distribution bus

The current drawn by a real IC, or other load, would usually depend on the voltage V_L applied to its power pins. Here, for simplicity, we assume a constant load current I_L. This is the worst case, since the load does not then absorb any part of the transient. Both current sources dissipate power instead of generating it. We will discuss the voltage source V_2 later; until then, we assume it to be zero. The series resistance R_B is the DC resistance of the power-bus conductor. To avoid wasting power, we must use a bus conductor heavy enough that R_B is negligible, so we assume this resistance to be zero. The distributed inductance L_B and capacitance C_B, however, cannot be reduced so easily. They cause the power bus to behave as a transmission line.

A sudden change in I_T generates a transient that starts to discharge the capacitance C_B. Inductance L_B, however, prevents C_B from discharging instantaneously. The capacitance nearest the load discharges first, and as the remaining capacitance discharges, the transient propagates to the left along the bus, toward the power source V_S. When the transient reaches the source, the source starts to recharge C_B, and the transient begins to travel forward toward the load. Until the transient again reaches the load, the load voltage V_L differs from its steady-state (DC) value. The process continues for several equal intervals, until the transient dies out. Our greatest concern, however, is during the first interval, when the transient is strongest.

Assume that I_T is zero and has been so for a long time. Then the current on the power bus is steady DC, equal to I_L. There is no voltage across the series inductance and no current through the shunt capacitance. Since we assume R_B to be zero, there is no voltage across it either, and $V_L = V_S$. Now let I_T suddenly change from zero to some nonzero value, corresponding to a change in the current drawn by the IC or other load. Since the power bus is linear, we may superimpose the current I_T onto the bus and analyze it separately from the DC current I_L that is already there. The transient current I_T cannot flow through the current source I_L, so it propagates to the left on the bus, as a step function. The current I_T causes a voltage V_T, also a step function, to propagate along the bus with it. The positive pole of V_T is at the terminal of the current source from which I_T is flowing outward, which is the bottom terminal in Figure 6-10. The ratio V_T/I_T is constant and equal to the *characteristic impedance* Z_0 of the bus.[1,2]

When the transient reaches V_S, a reflected transient of opposite voltage polarity starts to propagate to the right, toward the load. Until this reflected transient reaches the load, the additional voltage V_T, equal to $I_T Z_0$, appears across the load. Thus, V_T is present at the load for the full round-trip transition interval. The voltage V_T adds to V_L but is of the opposite polarity, so V_L drops by the amount V_T during the transition interval. With the ideal voltage and current sources shown here, the

---

1. C. W. Davidson, *Transmission Lines for Communications,* 2nd ed. (New York, N.Y.: John Wiley & Sons, Inc., 1989), pp. 19–23.

2. John D. Kraus, *Electromagnetics,* 4th ed. (New York, N.Y.: McGraw-Hill Publishing Company, 1992), pp. 527–30.

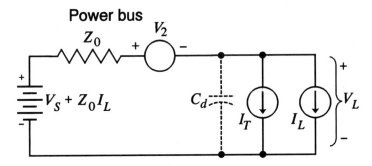

Figure 6-11   Thévenin equivalent of power bus

reflections would continue forever. Real circuits, however, contain resistive loads, so the transient will die out.

During the first transient interval, which is the interval of our concern, we may represent the power bus by the Thévenin equivalent circuit of Figure 6-11.[3] This is *not* the actual circuit, since Z_0 is not a physical resistance. However, we will now show that, at the load, the power bus *behaves the same as* Figure 6-11 *during the first transient interval only.* For a steady DC load current, where $I_T = 0$, the voltage across the series resistance Z_0 is $I_L Z_0$, and then $V_L = V_S$, as shown above. During a transient, $I_T \neq 0$, and the voltage across the series resistance is $(I_L + I_T)Z_0$. Then

$$V_L = V_S - I_T Z_0 = V_S - V_T \tag{6-11}$$

This is the correct value for V_L during the first round trip of the transient, also shown above. This Thévenin equivalent circuit is therefore usable at the load during the first transient interval. We will use this equivalent circuit in the analyses to follow.

Noise also can be capacitively or inductively coupled into the power bus from a nearby circuit. Inductive coupling is more likely since the load impedance is usually low for a real load such as a digital integrated circuit (IC). The series voltage source V_2 in Figures 6-10 and 6-11 represents inductively coupled noise, as discussed in Chapter 3. Either source, V_2 or I_T, will cause an AC voltage to appear across the load. The voltage supplied to the load is then no longer pure DC, and this may affect circuit operation. If the load is an emitter-follower amplifier, for example, part of the load current is the collector current of the transistor. The circuit may become unstable if V_L is not held constant. Another hazard exists if the load consists of two or more independent circuits, as shown in Figure 6-12. Here, variations in V_L caused by one circuit may affect the other. This is an example of conducted interference within the same system.

Since $V_T = I_T Z_0$ we can reduce V_T, and thus make V_L more constant, by lowering Z_0. Reducing Z_0 also inherently reduces the noise V_2 coupled from other circuits. If V_T is the maximum permissible variation in the supply voltage to an IC, and if I_T is the maximum variation in the load current, then the maximum permis-

---

3. Davidson, *Transmission Lines*, pp. 34–40.

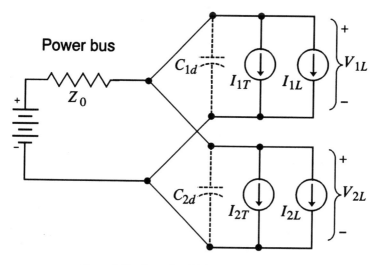

Figure 6-12   Power bus driving two varying loads

sible Z_0 for the power bus is V_I/I_T. The characteristic impedance of the power bus, or any transmission line,[4,5] is

$$Z_0 = \sqrt{\frac{L/l}{C/l}} \tag{6-12}$$

where L/l and C/l are the distributed inductance and capacitance per unit length, as discussed in Chapter 3. For a power bus consisting of two cylindrical conductors, we may use the formulae of Chapter 3 to calculate L/l and C/l, and from these, Z_0. The lowest characteristic impedance, however, occurs with a power bus consisting of two parallel flat conductors configured as a stripline. Neglecting field fringing at its edges, the capacitance of a stripline is

$$C = \frac{Q}{V} = \frac{|D|lw}{|E|t} = \frac{\epsilon lw}{t} \tag{6-13}$$

and thus the capacitance per unit length is

$$\frac{C}{l} = \frac{\epsilon w}{t} = \frac{\epsilon_r \epsilon_0 w}{t} \tag{6-14}$$

where w is the width and t is the thickness of the stripline, and ϵ_r is the relative permittivity of its dielectric. For a nonferrous dielectric, the inductance of a stripline is

$$L = \frac{\Psi}{I} = \frac{|B|lt}{|K|w} = \frac{\mu_0 |H|lt}{|H|w} = \frac{\mu_0 lt}{w} \tag{6-15}$$

4. Davidson, *Transmission Lines*, p. 22.

5. Kraus, *Electromagnetics*, pp. 488–92.

again neglecting field fringing. Thus the inductance per unit length is

$$\frac{L}{l} = \frac{\mu_0 t}{w} \tag{6-16}$$

The characteristic impedance of a stripline is therefore

$$Z_0 = \sqrt{\frac{L/l}{C/l}} = \sqrt{\frac{\mu_0 t/w}{\epsilon_r \epsilon_0 w/t}} = \frac{1}{\sqrt{\epsilon_r}} \sqrt{\frac{\mu_0}{\epsilon_0}} \left(\frac{t}{w}\right) = \frac{120\pi}{\sqrt{\epsilon_r}} \left(\frac{t}{w}\right) \Omega \tag{6-17}$$

For example, if $t = 1$ mm, $w = 2$ cm, and $\epsilon_r = 4$, then

$$Z_0 = \frac{120\pi}{2} \cdot \frac{0.1}{2} \ \Omega = 9.42 \ \Omega \tag{6-18}$$

A thinner dielectric would yield a proportionally lower characteristic impedance. If we know the value of Z_0 required to restrict V_T to an acceptable value, we can calculate the dimensions of the power bus using Equation (6-17).

If the PWC contains a ground plane, we may emulate the stripline by a single conductor on the PWC, usually as wide as is practical. The image of the power conductor, in the ground plane, forms the other stripline conductor. If there is insufficient space on the PWC for the power bus, then we may use a surface-mounted busbar or a multilayered PWC. The extreme is to use a *power plane* that covers the entire PWC, in the same way as a second ground plane. The power and ground planes may then be tapped at any point on the PWC, via plated-through holes.

The characteristic impedance of the combination of power plane and ground plane is very low. It depends on dielectric thickness and the distance between the power and ground pins of each IC. Even with a conventional stripline, the power pins cause additional series inductance and thus increase the power source impedance at the point of application to the load. Often, for large transients, the source impedance cannot be made low enough to provide a sufficiently constant load voltage V_L. Then, we can further stabilize V_L by adding the decoupling capacitor labeled C_d in Figure 6-11. Some of the transient current I_T then flows through C_d instead of propagating along the power bus. The amount of current depends on the ratio of Z_0 to the impedance of C_d. If the impedance of C_d is much less than Z_0, then most of I_T flows through C_d and the transient voltage on the line is greatly reduced. The impedance of C_d should therefore be as low as possible. There are practical limitations, of course, and we will now calculate the capacitor size necessary to keep V_T below a given limit. The method also will show whether the decoupling capacitor is necessary at all.

With C_d present, the current source I_T must discharge it and the distributed capacitance of the power bus. For this analysis we assume I_T to be a step function that is zero until the time $t = 0$ and constant after $t = 0$. The node equation for the V_L node in Figure 6-11, with $V_2 = 0$, is

$$\frac{V_S + Z_0 I_L - V_L}{Z_0} - C_d \frac{dV_L}{dt} - I_T - I_L = 0$$

$$C_d \frac{dV_L}{dt} + \frac{V_L}{Z_0} = \frac{V_S}{Z_0} - I_T \qquad (6\text{-}19)$$

With I_T constant after the time $t = 0$, the general solution for Equation (6-19) is the familiar exponential decay function

$$V_L = V_S - Z_0 I_T + K e^{-t/(Z_0 C_d)} \qquad (6\text{-}20)$$

where K is an arbitrary constant. The voltage V_L across C_d is equal to V_S until the time $t = 0$, and it cannot change instantaneously. Therefore, $V_L = V_S$ immediately after $t = 0$ also, and to satisfy Equation (6-20) for t very small, K must be equal to $Z_0 I_T$. The solution of Equation (6-19) that satisfies this initial condition is therefore

$$V_L = V_S - Z_0 I_T [1 - e^{-t/(Z_0 C_d)}] \qquad (6\text{-}21)$$

To be effective, the decoupling capacitor must prevent V_L from decreasing below the minimum acceptable power supply voltage, during the first round trip of the transient on the power bus. The propagation velocity v_p on the bus[6] is

$$v_p = \frac{1}{\sqrt{(L_B/l)(C_B/l)}} \qquad (6\text{-}22)$$

where L_B/l and C_B/l are the distributed inductance and capacitance of the power bus per unit length. If l is the physical length of the bus, the round-trip transit time T_t is

$$T_t = \frac{2l}{v_p} = 2l\sqrt{(L_B/l)(C_B/l)} = 2\sqrt{L_B C_B} \qquad (6\text{-}23)$$

where L_B and C_B are the total inductance and capacitance of the power bus. At the time T_t, the transient has returned to the load, and V_L has decreased to

$$V_L(T_t) = V_S - Z_0 I_T [1 - e^{-T_t/(Z_0 C_d)}]$$

$$= V_S - Z_0 I_T \left[1 - \exp\left(-\frac{2\sqrt{L_B C_B}}{Z_0 C_d} \right) \right]$$

$$= V_S - Z_0 I_T (1 - e^{-2C_B/C_d}) \qquad (6\text{-}24)$$

If V_T is the maximum permissible voltage variation at the IC power pin, $V_L(T_t)$ must be equal to $V_S - V_T$, or

$$V_S - V_T = V_S - Z_0 I_T (1 - e^{-2C_B/C_d}) \qquad (6\text{-}25)$$

Solving Equation (6-25) for C_d gives the required size of the decoupling capacitor,

---

6. Davidson, *Transmission Lines*, p. 8.

$$C_d = \frac{-2C_B}{\ln\left(1 - \dfrac{V_T}{Z_0 I_T}\right)} \qquad (6\text{-}26)$$

where I_T is the change in the load current, V_N is the maximum permissible voltage variation at the IC power pin, C_B is the total capacitance of the power bus, and Z_0 is its characteristic impedance. This formula assumes $V_T < I_T Z_0$; if this is not true, the decoupling capacitor is unnecessary. If much decoupling is necessary, that is, if $V_T \ll I_T Z_0$, then we may use the approximation $\ln(1 + x) \approx x$ for $x \approx 1$, and C_d is approximately given by

$$C_d \approx \frac{-2C_B}{-V_T/(Z_0 I_T)} = \frac{2C_B I_T Z_0}{V_T} \qquad (6\text{-}27)$$

Since $\ln(1 + x) \le x$ for every x, Equation (6-27) yields a value for C_d that is never less than that given by Equation (6-26). We may calculate all decoupling capacitors using this simpler expression, if desired.

The equivalent circuit of Figure 6-11 assumes no series inductance between I_T and C_d. For this assumption to be valid, the decoupling capacitor must be as close as possible to the power pin of the integrated circuit or other load.

A power bus usually drives multiple loads, as shown in Figure 6-12. The loads may be so far apart that the conductors between them behave as transmission lines. The above analysis then applies to each load. Let Z_0 and C_B now represent the characteristic impedance and total capacitance, respectively, of the portion of the bus that runs from the junction point to an individual load. A capacitor is necessary at each load for which V_T must be kept less than $I_T Z_0$. We must calculate C_d for each load, using the above values for Z_0 and C_B in Equation (6-26). If the junction point is some distance from the power source, another decoupling capacitor is necessary at the junction point. We calculate its value as above, with I_T now equal to the total transient current of all loads, and with Z_0 and C_B representing the characteristic impedance and total capacitance of the shared portion of the bus. If there are multiple distribution levels, a decoupling capacitor is advisable at each junction point, calculated as above.

If a perfect capacitor could be made, decoupling would be as simple as calculating C_d using Equation (6-26). Any practical capacitor, however, contains series resistance R_c and inductance L_c, and shunt conductance G_c. We now investigate the effects of these unavoidable parameters. Although they are distributed, we may approximate them by the lumped circuit in Figure 6-13. The impedance of the capacitor is

$$\begin{aligned}
Z_c &= j\omega L_c + R_c + \frac{1}{G_c + j\omega C_d} \\
&= \frac{(j\omega L_c + R_c)(G_c + j\omega C_d) + 1}{G_c + j\omega C_d} \\
&= \frac{1 - \omega^2 L_c C_d + R_c G_c + j\omega(G_c L_c + R_c C_d)}{G_c + j\omega C_d}
\end{aligned} \qquad (6\text{-}28)$$

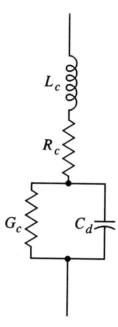

Figure 6-13 Equivalent circuit of a capacitor

The importance of L_c, R_c, and G_c depends on the operating frequency and the type of capacitor. Series resistance R_c is due to the finite conductivity of the metal deposited on the dielectric. Shunt conductance G_c is due to the finite resistivity of the dielectric. These parameters are normally significant only with electrolytic capacitors. For other capacitor types, R_c and G_c can be neglected, and Figure 6-12 becomes a simple series-resonant circuit. Equation (6-28) then reduces to

$$Z_c = \frac{1 - \omega^2 L_c C_d}{j\omega C_d} \qquad (6\text{-}29)$$

Then the major problem is due to the series inductance L_c, which depends partly on the length of the capacitor leads. To minimize L_c we must keep these leads as short as possible.

If we mount a decoupling capacitor directly between the power and ground pins of an IC, for example, the capacitor leads form a tiny current loop. The series inductance L_c can be no less than the inductance of this loop. From Chapter 4, the inductance of a circular loop is

$$L = \mu r \left(\ln \frac{164}{d} - 2 \right) \qquad (6\text{-}30)$$

where r is the radius of the loop and d is the diameter of the wire leads. As before, we estimate L_c by setting the loop size equal to the approximate size of the (non-circular) loop in the real circuit. For a 14-pin IC, the dimensions might be $r = 5$ mm, and $d = 0.64$ mm (#22 AWG). The approximate inductance is then

$$L_c = 2 \cdot 1.257 \ \mu H \cdot 0.005 \left(\ln \frac{16 \cdot 0.005}{0.64 \cdot 10^{-3}} - 2 \right)$$

$$= 0.01257 \cdot (\ln 125 - 2) \ \mu H$$

$$= 0.01257 \cdot 2.83 \ \mu H$$

$$= 0.0355 \ \mu H \tag{6-31}$$

Obviously for $\omega = 1/\sqrt{L_c C_d}$ (resonance), $Z_c = 0$, which is ideal since we are trying to achieve the lowest possible impedance. For the above dimensions, with $C_d = 0.047 \ \mu F$, the resonant frequency f_0 is

$$f_0 = \frac{\omega_0}{2\pi} = \frac{1}{2\pi \sqrt{L_c C_d}} = \frac{1}{2\pi \sqrt{0.0355 \ \mu H \cdot 0.047 \ \mu F}} = 3.90 \ \text{MHz} \tag{6-32}$$

Above this frequency, however, $|Z_c|$ *increases* with frequency and the capacitor has the opposite of its intended effect. Thus, a capacitor is less effective above its resonant frequency.

Even if its leads are very short, there is some inductance in the body of the capacitor itself. Therefore, the resonant frequency also depends on the type of capacitor. For polystyrene, mica, and some types of ceramic capacitors, the resonant frequency is above 100 MHz, neglecting the lead inductance. In these capacitor types, only small capacitance values are available. They are adequate at very high frequencies if the current loop at the IC pins (or other load) can be made small enough. For load currents that vary at lower frequencies, other capacitor types are suitable. Higher capacitance values are then available, but they also contain more inductance.

If the load current varies over a wide frequency band, a single capacitor may have too little capacitance at low frequencies or too much inductance at high frequencies. A solution to this difficulty is to connect two capacitors of different values, and different types, in parallel. The susceptance B_c of the combination is

$$B_c = \frac{\omega C_{d1}}{1 - \omega^2 L_{c1} C_{d1}} + \frac{\omega C_{d2}}{1 - \omega^2 L_{c2} C_{d2}}$$

$$= \frac{\omega C_{d1}(1 - \omega^2 L_{c2} C_{d2}) + \omega C_{d2}(1 - \omega^2 L_{c1} C_{d1})}{1 - \omega^2 (L_{c1} C_{d1} + L_{c2} C_{d2}) + \omega^4 L_{c1} C_{d1} L_{c2} C_{d2}}$$

$$= \frac{\omega(C_{d1} + C_{d2}) - \omega^3 C_{d1} C_{d2}(L_{c1} + L_{c2})}{1 - \omega^2 (L_{c1} C_{d1} + L_{c2} C_{d2}) + \omega^4 L_{c1} C_{d1} L_{c2} C_{d2}} \tag{6-33}$$

Unfortunately, the numerator of Equation (6-33) becomes zero at the frequency f_p, where

$$\omega_p^2 = \frac{C_{d1} + C_{d2}}{C_{d1}C_{d2}(L_{c1} + L_{c2})} = \frac{1/C_{d1} + 1/C_{d2}}{L_{c1} + L_{c2}}$$

$$f_p = \frac{\omega}{2\pi} = \frac{1}{2\pi}\sqrt{\frac{1/C_{d1} + 1/C_{d2}}{L_{c1} + L_{c2}}}$$

(6-34)

Unless $L_{c1}C_{d1} = L_{c2}C_{d2}$, which would imply identical resonant frequencies, the denominator of Equation (6-33) is not zero. Therefore, at the frequency f_p, $B_c = 0$ and the parallel combination has no effect. For unequal resonant frequencies, this parallel resonance is unavoidable. If the AC load current contains a frequency component near f_p, the capacitors will be ineffective at that frequency. The resonant frequency f_p of two decoupling capacitors in parallel must be a frequency at which the load does not generate any noise.

In extreme cases, decoupling capacitors may not provide enough suppression at one or more frequencies. Then, an individual ell-network filter may be necessary on each load, as shown for two loads in Figure 6-14. The series resistance does not reduce the transient voltage on the same load that causes it, but it does impede the transient from propagating to other loads. With respect to noise coupled from I_{1L} to R_{2L} and vice versa, the two ell networks, with C_d, form a low-pass RC ladder filter. This is easier to visualize if we redraw the circuit as shown in Figure 6-15.

A transient analysis of a ladder filter involves a higher-order differential equation, which we will not pursue here. Instead, we will consider the frequency content of the varying load current, I_T, and use a steady-state analysis. If $C_d \gg C_{1d}$, then the cutoff (half-power) frequency f_p of network 1 is

$$f_p = \frac{1}{2\pi R_{1d}C_{1d}}$$

(6-35)

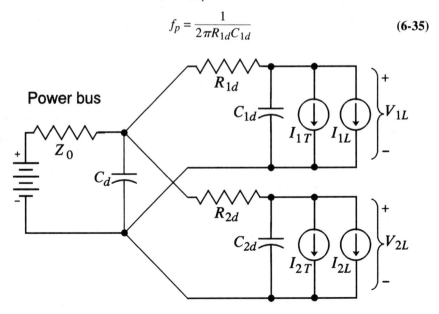

Figure 6-14   Power bus with ell network filter on each load

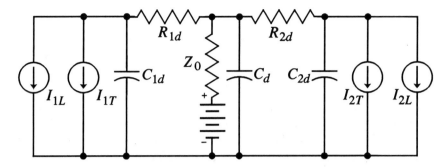

Figure 6-15   Circuit of Fig. 6-14 redrawn to emphasize noise path

and similarly for network 2. Above f_p, the noise voltage attenuation, in each section of the resistive ladder network, is proportional to frequency. Each series resistor R_{1d} and R_{2d} must be large enough so that f_p is well below the lowest frequency component of I_T. However, their resistance is limited by the allowable voltage drop. If the current is high, it may be necessary to use inductors instead. If L_{1d} is the inductance replacing R_{1d}, then the resonant frequency f_p of network 1 is equal to

$$f_p = \frac{1}{2\pi\sqrt{L_{1d}C_{1d}}} \qquad (6\text{-}36)$$

and similarly for network 2. Above f_p, the noise voltage attenuation, in each section of the inductive ladder network, is proportional to frequency squared. As with an RC filter, f_p must be well below the lowest frequency component of I_T. An LC filter, however, will be ineffective at f_p and will even *increase* the coupled noise at that frequency unless it is sufficiently damped. It is sufficiently damped if the series resistance R_{s1} of inductor L_{1d} satisfies the inequality

$$R > \sqrt{\frac{L_{1d}}{C_{1d}}} \qquad (6\text{-}37)$$

and similarly for inductor L_{2d}. To avoid resonance problems, the resistances of the inductors should satisfy Equation (6-37) so that the filter is sufficiently damped.

Besides the resonance problem, an inductive filter is more likely to radiate than is a resistor, and therefore the inductor must often be shielded. For these reasons, a resistor is usually preferable when the load current is moderate. A ferrite bead may be usable if only a small inductance is necessary. Ferrite beads exhibit hysteresis losses at higher frequencies, and the losses appear as a series resistance, which increases the damping. Because they do not increase the magnetic field outside the ferrite material, they do not increase the radiation. They will be discussed further in Chapter 11.

6-5 PROBLEMS

1. To which of the five classes of circuit does each of the following devices belong?
   (a) Public-address amplifier
   (b) Personal computer processor unit (CPU)

(c) Heat-generating circuitry of a microwave oven

(d) Timing circuitry of a microwave oven

(e) Telephone handset containing only a transmitter and receiver

(f) Rotary telephone dial

(g) TOUCH TONE® telephone dial

(h) TV tuner

(i) Electronic lamp dimmer

(j) Electronic organ (musical instrument)

2. Between which two classes of circuit would each of the following devices normally be connected?

(a) Conventional analog modulator

(b) Pulsed code (PCM) modulator

(c) Driver circuit for a computer-controlled motor

3. A PWC must contain all of the following circuits, each occupying its given size and shape.

(a) Audio amplifier, 2 in. × 3 in.

(b) Modulator, $2^1/_2$ in. × $2^1/_2$ in.

(c) Keypad, $2^1/_2$ in. × 3 in.

(d) High-frequency power converter, 2 in. × $2^1/_2$ in.

Arrange them on a rectangular board such that any two circuits may be interconnected without running any leads alongside a circuit of another class. It is acceptable to pass near a corner of another circuit.

4. In the first example in Section 6-3, the potential between the conductors of the first stripline is 10 V. Use the two-dimensional analogy between electric and magnetic fields to find the voltage that is capacitively coupled into the second stripline. Assume the same approximations as those used in the text for magnetic fields.

5. In the second example in Section 6-3, the capacitance between the conductors of the first stripline is 20 pF/m, and the voltage between them is 10 V. Find the electric field strength at the point where the magnetic field was determined, assuming similar approximations.

6. On a two-sided (non-multilayered) PWC, the lower side is intended to be a solid ground plane. The layout specialist has exhausted all other possibilities and says that he must include a stitch. However, the stitch could be avoided by using a wire jumper. With regard to electromagnetically coupled noise, why is a jumper preferable to a stitch?

7. A load is connected to the end of a power distribution bus, and the Thévenin equivalent is the circuit shown in Figure 6-11. The length of the bus is such that a transient generated at the load requires 5 ns to travel to the source and be reflected back to the load. Let $Z_0 = 1$ ohm, and let I_T be a step function equal to 1 ampere. The maximum change in V_L occurs just before the reflected wave returns to the load.

(a) What is the maximum change in V_L if C_d is not present?

(b) What is the maximum change in V_L if $C_d = 0.01$ μF?

(c) What value would be required for C_d in order to limit the change in V_L to -0.1 volts?

8. A certain 0.1-μF capacitor has a series resonant frequency of 2 MHz, and a 0.001-μF capacitor has a series resonant frequency of 50 MHz. If the two capacitors are connected in parallel, what is their parallel resonant frequency?

7

FREQUENCY SPECTRA
OF UNINTENTIONAL
RADIATION SOURCES

We have seen, in our examples of coupled noise, that frequency is nearly always a critical variable. Obtaining meaningful data requires some knowledge of the frequency components contained in the noise source. For intentional sources, this is well known; it is often a single radio frequency. These cases are comparatively easy to analyze. To use an equation that contains a function of f or ω, we simply substitute its known value.

Unintentional sources usually emit many frequencies. The frequency content depends on the type of circuit and the speed of its components. There may be only a few strong emissions at discrete frequencies, or there may be weak emissions spread over a large bandwidth. In this chapter, we shall study the characteristic frequencies of various equipment types and describe how to specify these emissions with respect to frequency.

7-1 TYPICAL WAVEFORMS

A voltage or current waveform may theoretically be any bounded function of time. A sine wave is a very common example, and because it is so common, it is the foundation of our analysis of coupled noise. Fourier analysis allows us to represent any bounded function as a finite or infinite summation of sine waves. The sine waves will have various amplitudes, frequencies, and phases. We assume in this text that the noise coupling process is linear, unless stated otherwise. Under this condition, we can treat each sine wave in the summation individually.

Some noise sources generate a steady, periodic waveform. Such noise sources are comparatively easy to analyze. Any physically realizable periodic waveform is representable by a sum of discrete frequencies that is usually infinite but may be finite. We know these as *harmonics* of the fundamental frequency, where the fundamental has the same period as the original waveform. Together they form the *Fourier series* expansion of the waveform. Let $v_p(t)$ be a periodic function of period T, f_n the nth harmonic frequency of $v_p(t)$, and k an arbitrary constant. Then the peak amplitude of each harmonic is the corresponding coefficient of the amplitude-phase form of the Fourier series. This, in turn, is twice the magnitude of the corresponding coefficient of the complex Fourier series, which is more convenient for the following analysis. The peak harmonic amplitude is expressible as a function $A(f_n)$ of the harmonic frequency f_n given by

$$A(f_n) = 2|V(f_n)| \text{ for } f_n > 0$$

$$A(0) = V(0)$$

$$\text{where } V(f_n) = \frac{1}{T} \int_k^{k+T} v_p(t) e^{-j2\pi f_n t} \, dt$$

(7-1)

Since $v_p(t)$ is periodic with period T, the integrals in Equation (7-1) are independent of k. If we include $A(f_n)$ as a parameter in a noise coupling expression, we account for all components. The total coupled noise is then simply the sum of these components. The only difficulty may be the intricacy of the mathematical functions.

Some noise sources, however, generate nonperiodic waveforms. We know that the frequency content of a nonperiodic waveform can be represented by a *Fourier transform*. A Fourier transform is a function of frequency that is comparable to the discrete Fourier series. It may contain impulses at a finite number of frequencies but is otherwise continuous. It is not directly applicable to coupled noise, however, since it assumes a known function of time. Moreover, if the noise function is not periodic, it must approach zero as $t \to \pm\infty$ if we are to evaluate its transform integral. Real noise sources do not behave this way. They continue indefinitely even if they are not periodic, and the waveforms are not exactly known. Instead, different functions occur at random. Such functions are called *sample functions*, and together all possible sample functions for a given noise source form a *random process*. Only some statistical information is known about the sample functions, so we must use statistical methods to analyze them. The reader should have a fundamental knowledge of probability and statistics and, in particular, random processes.

A random process is a function of time t and one or more random variables. It may be several different mathematical functions, each with its own probability of occurrence. The composite function may or may not have a fixed period with respect to t; if it does, the process is said to be periodic. An example of a random process is

$$v_r(t) = a \cos(2\pi f t + \phi)$$

(7-2)

where a, f, and ϕ are not known constants but statistically independent random variables. We must specify their probability densities to complete the definition of the random process. For this example, let the probability densities be

$$p_a(a) = \begin{cases} \dfrac{a}{\sigma_a^2} \exp\left(-\dfrac{a^2}{2\sigma_a^2}\right) & \text{for } a > 0 \\ 0 & \text{for } a \leq 0 \end{cases} \tag{7-3}$$

$$p_f(f) = \frac{1}{\sqrt{2\pi}\,\sigma_f} \exp\left[-\frac{(f-f_0)^2}{2\sigma_f^2}\right] \tag{7-4}$$

$$p_\phi(\phi) = \begin{cases} \dfrac{1}{2\pi} & \text{for } 0 \leq \phi < 2\pi \\ 0 & \text{otherwise} \end{cases} \tag{7-5}$$

where $\sigma_a = 1$ volt, $f_0 = 10$ MHz, and $\sigma_f = 2$ MHz.

If we knew the sample values for a, f, and ϕ, we would know the particular sample function of t. But since only their probability distributions are given, we do not know the exact function of t. Since the frequency f is not a fixed constant, neither is the period, and this process is not periodic.

Even for a periodic function, although the waveform within each period is known, the time origin is usually arbitrary. The time origin affects only the phase of each harmonic and not its amplitude or frequency. The coupled noise, in turn, usually depends only on the amplitude and frequency of each harmonic, not its phase. Thus, the coupled noise amplitude is usually independent of the time origin. A periodic function with its time origin unknown is a random process. An example is the process defined in Equation (7-2) with a and f replaced by known constants.

7-2 RESULTING BANDWIDTHS

A noise source, whether periodic or random, can affect a receiver if it emits any frequencies to which the receiver is sensitive. We will now examine the frequency content of the noise, and analyze how it may affect an intentional or unintentional receiver. This is known as *spectral analysis* of the noise.

A concept applicable both to periodic waveforms and random processes is the average *power* present at every frequency. Average power depends on amplitude but not phase. We assume a fictitious load resistance of 1 ohm. The average power at a given frequency is then the mean square of the voltage component at that frequency. For a sine wave of frequency f_n and amplitude $A(f_n)$, such as the nth harmonic of $v(t)$ in Equation (7-1), the average power is

$$P(f_n) = \frac{A^2(f_n)}{2} = 2|V(f_n)|^2 \tag{7-6}$$

where $V(f_n)$ is as defined in Equation (7-1). The units of $P(f_n)$ in Equation (7-6) are volts^2, which, if divided by the fictitious load resistance of 1 ohm, would become watts. Because the true load resistance is usually not 1 ohm, we do not perform the division, but instead keep the units as volts^2. The original time function could be a current instead of a voltage, in which case the units of power would be amperes^2. If the original time function is an electric field $E(t)$, we define a fictitious average power per square meter to be the mean square of the field intensity. The units for the electric field are volts/meter, so the fictitious power per square meter must have the units of volts^2/meter^2. It is related to the true power in watts/meter^2 by a factor known as the wave impedance, to be discussed in Chapter 8.

To simplify calculations, we may define a reference level and express the actual levels in decibels above this reference. The common reference levels for electromagnetics are 1 μV for voltages and 1 μV/m for electric fields, both denoting peak values. By setting $A(f_n)$ in Equation (7-6) to each of these voltage levels, we note that the corresponding average power levels are 0.5 μV^2 and 0.5 μV^2/m^2, respectively. When expressing voltages and field strengths with respect to these levels, we use the units dB$_{\mu V}$ and dB$_{\mu V/m}$, respectively. For a sine wave voltage of peak amplitude A and average power P, we have

$$V_{dB\mu V} = 20 \log \frac{A}{1 \mu V} = 10 \log \frac{P}{0.5 \mu V^2} \tag{7-7}$$

and for a sinusoidally varying field of peak amplitude E,

$$E_{dB\mu V/m} = 20 \log \frac{|E|}{1 \mu V/m} = 10 \log \frac{P}{0.5 (\mu V/m)^2} \tag{7-8}$$

For example, a peak field strength of 30 μV/m is equal to

$$E_{dB\mu V/m} = 20 \log \frac{30 \mu V/m}{1 \mu V/m} = 29.5 \ dB_{\mu V/m} \tag{7-9}$$

From Equation (7-1), since $v(t)$ is a real function of time,

$$P(f_n) = 2|V(f_n)|^2 = 2V * (f_n)V(f_n)$$

$$= \frac{2}{T^2} \left[\int_{k_1}^{k_1+T} v(t_1)e^{j2\pi f_n t_1} \, dt_1 \right]\left[\int_{k_2}^{k_2+T} v(t_2)e^{-j2\pi f_n t_2} \, dt_2 \right]$$

$$= \frac{2}{T^2} \int_{k_1}^{k_1+T}\int_{k_2}^{k_2+T} v(t_1)v(t_2)e^{-j2\pi f_n(t_2-t_1)}dt_2 dt_1 \tag{7-10}$$

With the change of variable $\tau = t_2 - t_1$, and with k_2 set equal to $k_1 + t_1$, Equation (7-10) becomes

$$P(f_n) = \frac{2}{T^2} \left[\int_{k_1}^{k_1+T}\int_{k_2-t_1}^{k_2-t_1+T} v(t_1)v(t_1 + \tau)e^{-j2\pi f_n \tau} \, d\tau dt_1 \right.$$

$$= \frac{2}{T^2} \int_{k_1}^{k_1+T} \left[\frac{1}{T}\int_{k_1}^{k_1+T} v(t_1)v(t_1 + \tau)dt_1 \right] e^{-j2\pi f_n \tau} \, d\tau \tag{7-11}$$

The integral within the brackets of Equation (7-11) is the *autocorrelation function* of the periodic signal $v(t)$, designated $R(\tau)$.

$$R(\tau) = \frac{1}{T} \int_{k}^{k+T} v(t)v(t + \tau)\, dt \qquad (7\text{-}12)$$

The autocorrelation function is the correlation between the voltage $v(t)$ and a delayed or advanced version of it, $v(t + \tau)$. It depends on the amplitude and frequency of each harmonic of $v(t)$. It does not depend on the phase, since it is averaged over one period, which includes all phase angles. Although it is usually different from $v(t)$, $R(\tau)$ is periodic with period T. It therefore consists of discrete harmonics and is expressible as a Fourier series. Since $v(t)$ is periodic, Equation (7-12) may equivalently be written as follows:

$$R(\tau) = \frac{1}{T} \int_{k-\tau}^{k-\tau+T} v(t - \tau)v(t)\, dt$$

$$= \frac{1}{T} \int_{k}^{k+T} v(t)v(t - \tau)\, dt$$

$$= R(-\tau) \qquad (7\text{-}13)$$

Therefore, $R(\tau)$ is always an even function, and its complex Fourier series must contain only real terms.

From Equation (7-11) it follows that

$$P(f_n) = 2 \cdot \frac{1}{T} \int_{k}^{k+T} R(\tau)e^{-j2\pi f_n \tau}\, d\tau \qquad (7\text{-}14)$$

Thus, the average power contained in each frequency component of $v(t)$ is equal to twice the corresponding complex Fourier series coefficient of $R(\tau)$. This is not the series for $v(t)$ itself, but for the autocorrelation function of $v(t)$. If we know the correlation between $v(t)$ and $v(t + \tau)$ for every τ between 0 and T, then we know the power contained in each harmonic.

The effects of coupled noise depend on the bandwidth of the device that it affects (the victim). If the victim is an intentional receiver, for example, its bandwidth may be very narrow. Other devices, however, may be sensitive to a wide band of frequencies. If the bandwidth is great enough, several frequency components of the noise signal can simultaneously affect the circuit. Then the total average power affecting the victim is the sum of the average powers at the individual frequencies (see Problem 7). For example, if there are two components of equal amplitudes, the resulting average power is twice the average power in either component. More generally, consider a victim that is equally sensitive to noise at all frequencies between f_j and f_k but is immune to other noise frequencies. The total noise power $P(f_j, f_k)$ that affects the victim is then

$$P_{jk} = \sum_{i=j}^{k} P(f_i) \qquad (7\text{-}15)$$

To obtain the correct average power, we must compute the average over a full period of the resulting signal. If two component frequencies are nearly but not exactly equal, this period may become very long. The power averaged over a shorter interval may be greater or less than the true average power. Equation (7-6) does *not* apply to the total power and the total peak amplitude, for the signal is no longer sinusoidal. The total peak amplitude may be as great as the sum of the peak amplitudes of the components, depending on their relative frequencies and phases.

We now consider a random process instead of a known periodic function of time. Using statistical methods, we can find the correlation between the random values of the sample functions at two time instants. The autocorrelation function $R(t,\tau)$ of a random process is defined to be this correlation, at the two time instants t and $t + \tau$. The expression for $R(t,\tau)$ is

$$R(t, \tau) = \int_{-\infty}^{\infty} \int_{-\infty}^{\infty} \cdots \int_{-\infty}^{\infty} [p(x_1, x_2, \ldots, x_n) \cdot$$
$$v(t, x_1, x_2, \ldots, x_n)v(t + \tau, x_1, x_2, \ldots, x_n)]dx_n \ldots dx_2 dx_1 \qquad (7\text{-}16)$$

where x_1 through x_n are the random variables defining the process and $p(x_1, x_2, \ldots, x_n)$ is their joint probability density. The number of random variables may be very large or even infinite. For brevity, Equation (7-16) is sometimes written

$$R(t, \dot{\tau}) = \int_{-\infty}^{\infty} p(\mathbf{x})v(t, \mathbf{x})v(t + \tau, \mathbf{x})d\mathbf{x} \qquad (7\text{-}17)$$

Here, $\mathbf{x}$ is an n-dimensional vector representing x_1 through x_n, and $\int d\mathbf{x}$ is a multiple integral over all x's.

For the process defined in Equations (7-2) through (7-5), since $a, f,$ and ϕ are statistically independent, the autocorrelation function is

$$R(t, \tau) = \int_0^{\infty} \left\{ \int_{-\infty}^{\infty} \left\{ \int_0^{2\pi} p_a(a)p_f(f)p_\phi(\phi) \cdot \right. \right.$$
$$\left. \left. a^2 \cos(2\pi ft + \phi)\cos[2\pi f(t + \tau) + \phi]d\phi \right\}df \right\}da$$
$$= \left[\int_0^{\infty} p_a(a)a^2 da \right] \left\{ \int_{-\infty}^{\infty} p_f(f) \left[\int_0^{2\pi} p_\phi(\phi) \cdot \right. \right.$$
$$\left. \left. \frac{\cos(4\pi ft + 2\phi + 2\pi f\tau + \cos(2\pi f\tau)}{2}d\phi \right]df \right\}$$
$$= \left[\int_0^{\infty} p_a(a)a^2 da \right] \left\{ \int_{-\infty}^{\infty} \frac{p_f(f)}{2\pi} \cdot \right.$$
$$\left. \left[\frac{\sin(4\pi ft + 2\phi + 2\pi f\tau)}{4} + \frac{\phi \cos(2\pi f\tau)}{2} \right]_0^{2\pi} df \right\}$$
$$= \frac{1}{2} \left[\int_0^{\infty} p_a(a)a^2 da \right] \left[\int_{-\infty}^{\infty} p_f(f)\cos(2\pi f\tau)df \right]$$
$$= \sigma_a^2 \exp[-2(\pi \sigma_f \tau)^2]\cos(2\pi f_0 \tau) \qquad (7\text{-}18)$$

where $\sigma_a = 1$ volt, $f_0 = 10$ MHz, and $\sigma_f = 2$ MHz, as before. For this random process, the autocorrelation function does not depend on t, only τ. Such processes are said to be *stationary*. For a stationary process, the autocorrelation function is designated $R(\tau)$, and we again note that $R(-\tau) = R(\tau)$. We shall consider only stationary processes in this text.

The autocorrelation function of a periodic random process is periodic, just like the autocorrelation function of a known periodic function. We can represent it by a Fourier series, and each coefficient in the series represents the average power in the process at that frequency. Thus, a periodic random noise process behaves much like a known function of time and can be treated similarly.

The autocorrelation function of a nonperiodic random process is not periodic and therefore is not expressible as a discrete Fourier series. For a physically realizable noise process, however, the autocorrelation function does have a Fourier transform. This Fourier transform is called the *power spectral density* of the random process, designated $S(f)$. If the process contains no DC or periodic components, the transform is a continuous function of frequency. Its formula is

$$S(f) = \int_{-\infty}^{\infty} R(\tau)e^{-j2\pi f\tau}\,d\tau \qquad (7\text{-}19)$$

It does not represent the power present at the frequency f. That power must be zero, since the finite total power is spread continuously over an infinite number of points in the spectrum. Instead, the power contained in a small frequency band of width Δf, centered on the frequency f, is approximated by

$$P(f, \Delta f) \approx 2S(f)\,\Delta f \qquad (7\text{-}20)$$

In the limit as $\Delta f \to 0$, Equation (7-20) becomes exact. For a wider bandwidth, namely from f_j to f_k, the power contained in the band is

$$P(f_j, f_k) = 2\int_{f_j}^{f_k} S(f)\,df \qquad (7\text{-}21)$$

Compare Equation (7-21) with Equation (7-15), which gives the power in a given bandwidth for a periodic function or periodic random process.

For a voltage noise process, the units of $S(f)$ are volts^2/hertz. Some texts assume a 1-ohm load resistance and use watts/hertz. To avoid confusion when we later discuss radiated power, we will retain the units of volts^2/hertz when the load resistance is unknown. If the noise process were an electric field instead, its units would be volts/meter, so its power spectral density function would be expressed in volts^2/meter^2/hertz. As with a periodic noise source, we define a reference level for the spectral density and express the actual levels in decibels above it. For spectral densities the common reference levels are defined as 1 μV^2/MHz for voltages and 1 μV^2/m^2/MHz for fields. When expressing voltages and field strengths with respect to these levels, we abbreviate the units dB$_{\mu V/MHz}$ and dB$_{\mu V/m/MHz}$, respectively. Therefore, for a voltage random process,

$$V_{dB\mu V/MHz} = 10 \log \frac{P(f, \Delta f)/\Delta f}{1\ \mu V^2/MHz} = 10 \log \frac{2S(f)}{1\ \mu V^2/MHz} \qquad (7\text{-}22)$$

For an electric field process,

$$E_{\mathrm{dB}\mu\mathrm{V/m/MHz}} = 10 \log \frac{P(f, \Delta f)/\Delta f}{1\ \mu\mathrm{V}^2/\mathrm{m}^2/\mathrm{MHz}} = 10 \log \frac{2S(f)}{1\ \mu\mathrm{V}^2/\mathrm{m}^2/\mathrm{MHz}} \tag{7-23}$$

A noise source may contain both periodic and nonperiodic components. Therefore, it may simultaneously emit both types of noise. To deal with both types, we may generalize the Fourier transform to include the impulse function $\delta(f - f_i)$, where f_i is one of the discrete frequency components. Then periodic functions and processes have power spectral density functions containing impulses. They may be treated the same as nonperiodic processes if we follow the proper integration procedures. In particular, integration of $\delta(f - f_i)$ requires the sifting integral

$$\int_{x_1}^{x_2} f(x)\, \delta(x - x_0)\, dx = \begin{cases} f(x_0) & \text{for } x_1 < x_0 < x_2 \\ \tfrac{1}{2}f(x_0) & \text{for } x_0 = x_1 \text{ or } x_2 \\ 0 & \text{otherwise} \end{cases} \tag{7-24}$$

The frequency of a real signal source is never perfectly precise. Even the slightest uncertainty or variation in frequency causes the source to become a random process, strictly speaking. As the frequency becomes more precise, the power spectral density of this random process more nearly approaches the function $\delta(f - f_i)$. But since it never truly reaches $\delta(f - f_i)$, all practical sources are, strictly, nonperiodic random processes. To treat the two types of noise, we require a more practical distinction between them.

Practically, we may treat a noise source as periodic if its spectral density function contains only impulses or approximate impulses. But how close is "approximate"? Also, two approximate impulses may be at frequencies so near that they overlap. Should these be considered a nonperiodic source or two periodic sources?

To answer these questions, the following definition is applied. For measurement purposes, the receiver bandwidth must always be specified. If the bandwidth of a component of the noise source exceeds the bandwidth of the receiver, that noise component is *broadband*. The noise source is usually a nonperiodic random process but could be a periodic signal whose fundamental frequency is less than the receiver bandwidth. It behaves like a nonperiodic random process, so the received power depends on the receiver bandwidth, as shown in Equation (7-20). The power received from the random process defined in Equations (7-2) through (7-5), with a receiver bandwidth of 100 kHz, is as shown in Figure 7-1. This graph results from substituting the spectral density of this process into Equation (7-20), with $\Delta f = 100$ kHz. Since the 2-MHz bandwidth of the random process exceeds the receiver bandwidth, the noise is broadband. Reducing the receiver bandwidth would cause the peak power to decrease approximately in proportion to the bandwidth.

On the other hand, if the bandwidth of a noise component is less than the receiver bandwidth, then the noise component is *narrowband*. Such a noise source is usually periodic or *nearly periodic*, which means that adjacent intervals of its autocorrelation function differ only slightly. The power received from a narrowband noise component does not depend on the receiver bandwidth. Whenever a nar-

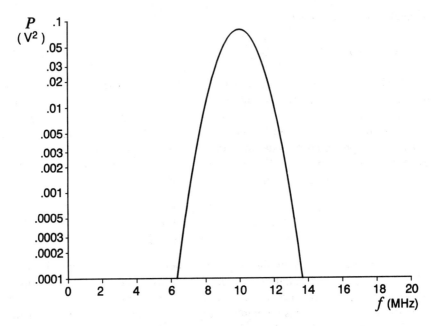

Figure 7-1 Broadband power spectrum; receiver bandwidth = 100 kHz

rowband noise component falls completely within the receiver pass band, the receiver will detect all power contained in that component, whatever its bandwidth. At the edges of the pass band, the received power will depend on the skirt selectivity of the receiver.

The power spectrum received from a 20-MHz, 1-V peak-to-peak sawtooth wave, with a receiver bandwidth of 100 kHz, appears in Figure 7-2. The sawtooth wave is a periodic function, so its power spectral density contains only impulses. Since an impulse has zero bandwidth, the noise is narrowband. The strength of each impulse is obtainable from Equations (7-1) and (7-6). The graph in Figure 7-2 includes the effects of the receiver bandwidth, and the width of each peak on the graph depends on the bandwidth. If the receiver bandwidth is reduced, the peaks become narrower but not shorter.

7-3 SUMMATION OF MULTIPLE NOISE SOURCES

We will need to consider cases in which noise accumulates from multiple sources. If the sources are at different frequencies, the total noise power is simply the sum of the powers in the individual noise components, as mentioned previously. If they include components at the same frequency, however, the summation process is more complicated.

Consider two noise sources that are sinusoidal functions of time, at the same frequency f, with known amplitudes and phases.

$$v_1(t) = A_1 \cos(2\pi ft + \phi_1) \qquad v_2(t) = A_2 \cos(2\pi ft + \phi_2) \qquad \text{(7-25)}$$

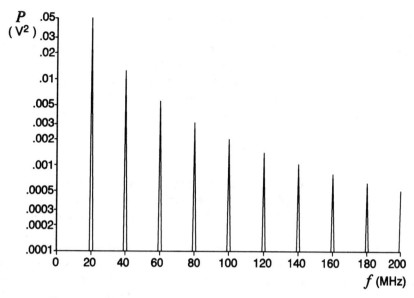

Figure 7-2   Narrowband power spectrum; receiver bandwidth = 100 kHz

From Equation (7-6), the average power in each of these signals is

$$P_1 = \frac{A_1^2}{2} \qquad P_2 = \frac{A_2^2}{2} \tag{7-26}$$

However, if we add the two voltages, the power in the combined signal is

$$P_{12} = \lim_{T \to \infty} \int_{-T/2}^{T/2} \frac{[v_1(t) + v_2(t)]^2}{T}\, dt$$

$$= \lim_{T \to \infty} \int_{-T/2}^{T/2} \frac{[A_1 \cos(2\pi ft + \phi_1) + A_2 \cos(2\pi ft + \phi_2)]^2}{T}\, dt$$

$$= \lim_{T \to \infty} \int_{-T/2}^{T/2} \left[\frac{A_1^2 \cos^2(2\pi ft + \phi_1)}{T} + \frac{A_2^2 \cos^2(2\pi ft + \phi_2)}{T} + \right.$$
$$\left. \frac{2A_1 A_2 \cos(2\pi ft + \phi_1) \cos(2\pi ft + \phi_2)}{T} \right] dt$$

$$= P_1 + P_2 + A_1 A_2 \lim_{T \to \infty} \int_{-T/2}^{T/2} \frac{\cos(4\pi ft + \phi_1 + \phi_2) + \cos(\phi_1 - \phi_2)}{T}\, dt$$

$$= P_1 + P_2 + A_1 A_2 \cos(\phi_1 - \phi_2) = P_1 + P_2 + 2\sqrt{P_1 P_2}\, \cos(\phi_1 - \phi_2) \tag{7-27}$$

Unless $|\phi_1 - \phi_2| = 90°$ or $270°$, this is not equal to the sum of the powers in the two individual signals. The difference in power does not violate conservation of energy. It simply means that for a given peak amplitude A, each source must provide more power in the presence of the other source than it would by itself.

From Equation (7-27) we note that the power depends on $\phi_1 - \phi_2$, the relative phase angle between the two sources. If their relative phase is constant, then the components add *coherently*. This would be true, for example, if the components are somehow synchronized. Depending on their relative phase, the total noise power may differ from the sum of the individual powers by as much as twice their geometric mean. If we do not know the relative phase $\phi_1 - \phi_2$, the only bound that we can establish for the total power is to assume that $\cos(\phi_1 - \phi_2) = 1$, or

$$P_{12} \le P_1 + P_2 + 2\sqrt{P_1 P_2} = (\sqrt{P_1} + \sqrt{P_2})^2 = \frac{(A_1 + A_2)^2}{2} \qquad (7\text{-}28)$$

The total power is thus no greater than the power in a signal whose amplitude is the sum of the individual amplitudes. If the two components are of equal amplitudes, the resulting noise power may be any value between zero and four times the power in either component by itself.

If two noise components are generated by independent, unsynchronized sources, then their relative phase varies randomly. In this case the components add *incoherently*. To calculate the expected power in the sum of the signals we must treat the sum as a random process. The phases ϕ_1 and ϕ_2 are random variables. For the phases to be completely random, they must be statistically independent and uniformly distributed between 0 and 2π. Their joint probability density function is therefore

$$p_\phi(\phi_1, \phi_2) = \begin{cases} \dfrac{1}{4\pi^2} & \text{for } 0 \le \phi_1, \phi_2 < 2\pi \\ 0 & \text{otherwise} \end{cases} \qquad (7\text{-}29)$$

The autocorrelation function of the sum is (see Problem 8)

$$R(\tau) = \frac{A_1^2 + A_2^2}{2} \cos(2\pi f\tau) \qquad (7\text{-}30)$$

The power at the frequency f is the Fourier series coefficient of the term for this frequency (the only term in the series):

$$P_{12} = \frac{A_1^2 + A_2^2}{2} \qquad (7\text{-}31)$$

We immediately note that P_{12} is the sum of the powers in the individual components. This is only the *expected* total power, a statistical estimate that is not very consistent when only two signals are added. When we later discuss the sum of many sources, however, the estimate will become more consistent.

We conclude that when two periodic signals add, the expected total power is the sum of the powers in the individual signals *unless* they are of the same frequency and their relative phases are constant. Then the power may be as great as that in a signal whose amplitude is the sum of the individual amplitudes. When two random processes add, periodic or not, they behave much like the two periodic functions of time discussed above. If there is no correlation between the processes, they add incoherently and the expected power in the sum is again equal to the sum of the individual powers. If the processes are correlated in some manner, the expected total power is usually different and may be higher. All random processes considered in this book will be uncorrelated, so we may assume that they add incoherently.

Our usual application of this result will involve the summation of many sources instead of only two. Extension of the summation technique to N sources is obvious. By mathematical induction we may show that with N sources instead of only two, the sum of the N terms replaces the two terms in Equations (7-28) and (7-31). If N identical noise sources add incoherently, each having the power P_0, then

$$P_{tot} = \sum_{i=1}^{N} P_0 = NP_0 \qquad (7\text{-}32)$$

This occurs if the sources are not synchronized. If they are synchronized, they add coherently, and

$$P_{tot} \leq \left(\sum_{i=1}^{N} \sqrt{P_0} \right)^2 = (N\sqrt{P_0})^2 = N^2 P_0 \qquad (7\text{-}33)$$

Thus the total noise power from N synchronized sources may be as great as N^2 times the power in each source. This can cause noise power to increase rapidly as the number of synchronized noise sources increases. As we will discuss in Chapter 8, from a noise standpoint it is best to minimize the number of synchronized circuits.

7-4 SPECTRUM OF A DIGITAL RANDOM PROCESS[1]

A particularly important random process is a random binary signal. Digital computing equipment contains many sources of these signals, with limited statistical information known about them. Let V_0 be the logic "1" voltage, and assume that the logic "0" voltage is 0 volts. The sample functions are random sequences of these voltage pulses, of arbitrary time duration.

Most digital systems are synchronous, controlled by a periodic time function commonly known as a clock. These are the systems that we will consider. In such a system, a logic level can change only at the instant of a clock pulse. Let f_c be the clock frequency. We may regard any signal in the system as a sequence of pulses, with the duration of each pulse equal to the clock period $T_c = 1/f_c$. Each pulse may or may not be present, depending on the digital data.

Practical circuits must have nonzero rise and fall times. We will assume equal rise and fall times and let t_r represent their duration. The leading edge may

1. William R. Bennett and James R. Davey, *Data Transmission* (New York, N.Y.: McGraw-Hill Publishing Company, 1965), pp. 315–21. Adapted with permission.

be any function of time g_r such that $g_r(0) = 0$ and $g_r(t_r) = V_0$. We will consider only the cases for which the trailing edge is the inverse of the leading edge, or $V_0 - g_r(t - T_c)$. Then, since the leading and trailing edges of two consecutive pulses overlap and their sum is equal to V_0, the pulses are equivalent to a single longer pulse. A signal that is equal to logic "1" for several clock periods is then representable by consecutive pulses. A signal with a duration of three clock periods is shown in Figure 7-3. After deriving a general expression for any such process, we will consider the specific case of trapezoidal pulses.

Let $g(t)$ be the function representing a single digital pulse corresponding to a clock period of T_c, beginning at $t = 0$. Mathematically,

$$g(t) = \begin{cases} g_r(t) & \text{for } 0 \le t < t_r \\ V_0 & \text{for } t_r \le t < T_c \\ V_0 - g_r(t - T_c) & \text{for } T_c \le t < T_c + t_r \\ 0 & \text{otherwise} \end{cases} \qquad (7\text{-}34)$$

Each sample function of the random process is then equal to the sum of an infinite number of these pulses, each of which may or may not be present, depending on the digital data. A sample function is given by

$$v(t) = \sum_{i = -\infty}^{+\infty} x_i \, g(t - iT_c - t_\phi) \qquad (7\text{-}35)$$

Here, each x_i is a binary random variable that assumes the value 0 or 1 according to the data value during interval i. The random variable t_ϕ represents the arbitrary time origin and varies from 0 to T_c with equal probability:

$$p(t_\phi) = \begin{cases} \dfrac{1}{T_c} & \text{for } 0 \le t_\phi < T_c \\ 0 & \text{otherwise} \end{cases} \qquad (7\text{-}36)$$

To define the random process, we must find the joint probability distribution of the x_i's. Knowing this information, we may then obtain the autocorrelation function of the process from Equation (7-17).

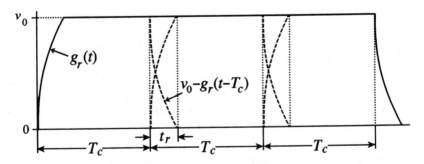

Figure 7-3   Digital signal with matched leading and trailing edges

The simplest case is to assume that the x_i's are statistically independent and that logic values of "0" and "1" are equally likely to occur. Then

$$p(\ldots, x_{-2}, x_{-1}, x_0, x_1, x_2, \ldots) = \prod_{k=-\infty}^{\infty} p(x_k) = \prod_{k=-\infty}^{\infty} \left[\frac{\delta(x_k)}{2} + \frac{\delta(x_k - 1)}{2} \right] \qquad (7\text{-}37)$$

Substituting Equations (7-35), (7-36), and (7-37) into Equation (7-17) then yields the autocorrelation function

$$R(t, \tau) = \int_{-\infty}^{\infty} \int_{-\infty}^{\infty} p(t_\phi) p(\mathbf{x}) v(t, \mathbf{x}, t_\phi) v(t + \tau, \mathbf{x}, t_\phi) \, d\mathbf{x} dt_\phi$$

$$= \int_0^{T_c} \int_{-\infty}^{\infty} \frac{1}{T_c} \prod_{k=-\infty}^{\infty} \left[\frac{\delta(x_k)}{2} + \frac{\delta(x_k - 1)}{2} \right] \cdot$$

$$\sum_{i=-\infty}^{+\infty} x_i g(t - iT_c - t_\phi) \sum_{j=-\infty}^{+\infty} x_j g(t + \tau - jT_c - t_\phi) \, d\mathbf{x} dt_\phi$$

$$= \sum_{i=-\infty}^{+\infty} \sum_{j=-\infty}^{+\infty} \left\{ \int_{-\infty}^{\infty} \prod_{k=-\infty}^{\infty} \left[\frac{\delta(x_k)}{2} + \frac{\delta(x_k - 1)}{2} \right] x_i x_j d\mathbf{x} \right\} \cdot$$

$$\int_0^{T_c} \frac{1}{T_c} g(t - iT_c - t_\phi) g(t + \tau - jT_c - t_\phi) \, dt_\phi \qquad (7\text{-}38)$$

The integral within the braces in Equation (7-38) is a multiple sifting integral. Like any multiple integral of a product of functions of distinct variables, it is equal to the product of the integrals of the individual functions. If $i \neq j$, it is equal to

$$\int_{-\infty}^{\infty} \prod_{k=-\infty}^{\infty} \left[\frac{\delta(x_k)}{2} + \frac{\delta(x_k - 1)}{2} \right] x_i x_j d\mathbf{x}$$

$$= \left\{ \prod_{\substack{k=-\infty \\ k \neq i,j}}^{\infty} \int_{-\infty}^{\infty} \left[\frac{\delta(x_k)}{2} + \frac{\delta(x_k - 1)}{2} \right] dx_k \right\} \cdot$$

$$\int_{-\infty}^{\infty} \left[\frac{\delta(x_i)}{2} + \frac{\delta(x_i - 1)}{2} \right] x_i dx_i \cdot \int_{-\infty}^{\infty} \left[\frac{\delta(x_j)}{2} + \frac{\delta(x_j - 1)}{2} \right] x_j dx_j$$

$$= \left[\prod_{\substack{k=-\infty \\ k \neq i,j}}^{\infty} \left(\frac{1}{2} + \frac{1}{2} \right) \right] \cdot \left(\frac{0}{2} + \frac{1}{2} \right) \cdot \left(\frac{0}{2} + \frac{1}{2} \right)$$

$$= \frac{1}{4} \qquad (7\text{-}39)$$

For $i = j$, the integral is instead equal to

$$\int_{-\infty}^{\infty} \prod_{k=-\infty}^{\infty} \left[\frac{\delta(x_k)}{2} + \frac{\delta(x_k - 1)}{2}\right] x_i x_j d\mathbf{x}$$

$$= \left\{ \prod_{\substack{k=-\infty \\ k \neq i}}^{\infty} \int_{-\infty}^{\infty} \left[\frac{\delta(x_k)}{2} + \frac{\delta(x_k - 1)}{2}\right] dx_k \right\} \cdot \int_{-\infty}^{\infty} \left[\frac{\delta(x_i)}{2} + \frac{\delta(x_i - 1)}{2}\right] x_i x_i dx_i$$

$$= \left[\prod_{\substack{k=-\infty \\ k \neq i}}^{\infty} \left(\frac{1}{2} + \frac{1}{2}\right) \right] \cdot \left(\frac{0 \cdot 0}{2} + \frac{1 \cdot 1}{2}\right)$$

$$= \frac{1}{2} \tag{7-40}$$

Substituting these values into Equation (7-38) then yields

$$R(t, \tau) = \sum_{i=-\infty}^{+\infty} \sum_{j=-\infty}^{+\infty} \frac{1}{4} \int_0^{T_c} \frac{1}{T_c} g(t - iT_c - t_\phi) g(t + \tau - jT_c - t_\phi) \, dt_\phi +$$

$$\sum_{i=-\infty}^{+\infty} \frac{1}{4} \int_0^{T_c} \frac{1}{T_c} g(t - iT_c - t_\phi) g(t + \tau - iT_c - t_\phi) \, dt_\phi$$

$$= \frac{1}{4} \int_0^{T_c} \frac{1}{T_c} \sum_{i=-\infty}^{\infty} g(t - iT_c - t_\phi) \sum_{j=-\infty}^{\infty} g(t + \tau - jT_c - t_\phi) \, dt_\phi +$$

$$\sum_{i=-\infty}^{+\infty} \frac{1}{4} \int_{iT_c}^{(i+1)T_c} \frac{1}{T_c} g(t - t_\phi') g(t + \tau - t_\phi') \, dt_\phi'$$

$$= \frac{1}{4} \int_0^{T_c} \frac{1}{T_c} \sum_{i=-\infty}^{\infty} g(t - iT_c - t_\phi) \sum_{j=-\infty}^{\infty} g(t + \tau - jT_c - t_\phi) \, dt_\phi +$$

$$\frac{1}{4} \int_{-\infty}^{\infty} \frac{1}{T_c} g(t - t_\phi') g(t + \tau - t_\phi') \, dt_\phi' \tag{7-41}$$

The summations $\sum_{i=-\infty}^{\infty} g(t - iT_c - t_\phi)$ and $\sum_{j=-\infty}^{\infty} g(t + \tau - jT_c - t_\phi)$ each include half of the result of Equation (7-40); therefore, the remaining half appears as a separate term with a coefficient of 1/4. These summations comprise a pulse sequence containing a pulse in every clock period, which is simply a steady DC signal equal to V_0. The first term of Equation (7-41) then reduces to a constant representing the DC component of the random process. With the change of variable $u = t - t'_\phi$ in the integral in its second term, Equation (7-41) becomes

$$R(\tau) = \frac{V_0^2}{4} + \frac{1}{4T_c} \int_{-\infty}^{\infty} g(u)g(u + \tau) \, du \qquad (7\text{-}42)$$

Since u is only a dummy variable of integration, t has now disappeared from the expression and R is a function only of τ. Thus the process is stationary, as expected.

The only remaining step necessary to evaluate $R(\tau)$ is to evaluate the integral in Equation (7-42), which of course depends on the function g. The integral must converge, since $g(t)$ is nonzero only over the finite range $0 \le t \le T_c + t_r$. The integration requires separate consideration of three different ranges of τ, namely, $0 \le \tau < t_r$, $t_r \le \tau < T_c - t_r$, and $T_c - t_r \le \tau < T_c + t_r$.

Since we are mainly interested in the spectral density, we may instead compute the Fourier transform of $R(\tau)$ in terms of the function g, as follows. From Equations (7-19) and (7-42), the spectral density of the random process is

$$S(f) = \int_{-\infty}^{\infty} \left[\frac{V_0^2}{4} + \frac{1}{4T_c} \int_{-\infty}^{\infty} g(u)g(u + \tau) \, du \right] e^{-j2\pi f\tau} \, d\tau$$

$$= \int_{-\infty}^{\infty} \frac{V_0^2}{4} e^{-j2\pi f\tau} \, d\tau + \frac{1}{4T_c} \int_{-\infty}^{\infty} \int_{-\infty}^{\infty} g(u)g(u + \tau) e^{-j2\pi f(u + \tau - u)} \, d\tau \, du$$

$$= \frac{V_0^2}{4} \delta(f) + \frac{1}{4T_c} \int_{-\infty}^{\infty} \int_{-\infty}^{\infty} g(u)g(u') e^{-j2\pi f(u' - u)} \, du' \, du$$

$$= \frac{V_0^2}{4} \delta(f) + \frac{1}{4T_c} \int_{-\infty}^{\infty} g(u') e^{-j2\pi fu'} \, du' \int_{-\infty}^{\infty} g(u) e^{j2\pi fu} \, du \qquad (7\text{-}43)$$

Let $G(f)$ represent the Fourier transform of $g(t)$, or

$$G(f) = \int_{-\infty}^{\infty} g(t) e^{-j2\pi ft} \, dt \qquad (7\text{-}44)$$

Then Equation (7-43) becomes

$$S(f) = \frac{V_0^2}{4} \delta(f) = \frac{1}{4T_c} G(f)G * (f) = \frac{V_0^2}{4} \delta(f) + \frac{f_c}{4} |G(f)|^2 \qquad (7\text{-}45)$$

The DC component of the random process, represented by the first term in Equation (7-45), is of no interest in electromagnetic compatibility. The second term represents the noise spectrum, which thus can be computed from the Fourier transform of a single pulse. This is usually simpler than computing the autocorrelation function.

Let $G_r(f)$ represent the Fourier transform of the leading edge function $g_r(t)$, and assume that $g_r(t)$ is zero outside the interval $0 \le t \le t_r$. Then $G(f)$ is given by

$$G(f) = \int_{-\infty}^{\infty} g(t)e^{-j2\pi ft}\, dt$$

$$= \int_0^{t_r} g_r(t)e^{-j2\pi ft}\, dt + \int_{t_r}^{T_c} V_0 e^{-j2\pi ft}\, dt + \int_{T_c}^{T_c+t_r} [V_0 - g_r(t-T_c)]e^{-j2\pi ft}\, dt$$

$$= \int_0^{t_r} g_r(t)e^{-j2\pi ft}\, dt + V_0 \frac{e^{-j2\pi ft_r} - e^{-j2\pi f(T_c+t_r)}}{j2\pi f} - \int_0^{t_r} g_r(t')e^{-j2\pi f(t'+T_c)}\, dt'$$

$$= (1 - e^{-j2\pi fT_c}) \left[\int_0^{t_r} g_r(t)e^{-j2\pi ft}\, dt + \frac{V_0 e^{-j2\pi ft_r}}{j2\pi f} \right]$$

$$= (1 - e^{-j2\pi fT_c}) \left[G_r(f) + \frac{V_0 e^{-j2\pi ft_r}}{j2\pi f} \right] \tag{7-46}$$

For a given function g_r, Equation (7-46) may be used to compute the Fourier transform of a single pulse, and from that the spectral density of the process.

We will now find the power spectral density of a process consisting of trapezoidal pulses, for which $g_r(t) = V_0 t/t_r$ for $0 \le t < t_r$. This is often used as an approximation of the pulse shapes found in actual circuits. The Fourier transform of a single pulse of the process is

$$G(f) = (1 - e^{-j2\pi fT_c}) \left[\int_0^{t_r} \frac{V_0 t}{t_r} e^{-j2\pi ft}\, dt + \frac{V_0 e^{-j2\pi ft_r}}{j2\pi f} \right]$$

$$= (1 - e^{-j2\pi fT_c}) \left[\frac{V_0}{t_r} \cdot \frac{(-j2\pi ft_r - 1)e^{-j2\pi ft_r} - (-1)}{(-j2\pi f)^2} + \frac{V_0 e^{-j2\pi ft_r}}{j2\pi f} \right]$$

$$= (1 - e^{-j2\pi fT_c}) \left[\frac{V_0}{t_r} \cdot \frac{1 - e^{-j2\pi ft_r}}{(-j2\pi f)^2} \right]$$

$$= \frac{V_0}{t_r} e^{-j2\pi f(T_c+t_r)} \frac{\sin(\pi fT_c)\sin(\pi ft_r)}{\pi^2 f^2} \tag{7-47}$$

From Equation (7-45) the spectral density of the process is

$$S(f) = \frac{V_0^2}{4} \delta(f) + \frac{f_c}{4} \left| \frac{V_0}{t_r} e^{-j\pi f(T_c+t_r)} \frac{\sin(\pi fT_c)\sin(\pi ft_r)}{\pi^2 f^2} \right|^2$$

$$= \frac{V_0^2}{4} \delta(f) + \frac{V_0^2 T_c}{4} \cdot \frac{\sin^2(\pi fT_c)}{\pi^2 f^2 T_c^2} \cdot \frac{\sin^2(\pi ft_r)}{\pi^2 f^2 t_r^2} \tag{7-48}$$

Then, for $f > 0$, the power detected by a receiver of bandwidth Δf is

$$P(f, \Delta f) = 2S(f)\,\Delta f = \frac{V_0^2 T_c \Delta f}{2} \cdot \frac{\sin^2(\pi fT_c)}{\pi^2 f^2 T_c^2} \cdot \frac{\sin^2(\pi ft_r)}{\pi^2 f^2 t_r^2} \tag{7-49}$$

A plot of this spectral density, for $T_c = 10t_r$, appears in Figure 7-4, where the 0-dB level corresponds to $V_0^2 T_c$. The DC component does not appear, since the logarithmic frequency scale does not include zero. The positions of the nulls depend on the rise time and clock frequency and are unimportant. What is significant is the *envelope* of the function, shown by the heavy line.

For every x, $\sin x \le 1$ and $(\sin x)/x \le 1$. Therefore, $S(f)$ must always satisfy the following three inequalities. First,

$$S(f) = \frac{V_0^2 T_c}{4} \cdot \frac{\sin^2(\pi f T_c)}{\pi^2 f^2 T_c^2} \cdot \frac{\sin^2(\pi f t_r)}{\pi^2 f^2 t_r^2}$$

$$\le \frac{V_0^2 T_c}{4} \cdot 1^2 \cdot 1^2 = \frac{V_0^2 T_c}{4} \qquad (7\text{-}50)$$

In addition,

$$S(f) = \frac{V_0^2}{4T_c \pi^2 f^2} \sin^2(\pi f T_c) \frac{\sin^2(\pi f t_r)}{\pi^2 f^2 t_r^2}$$

$$\le \frac{V_0^2}{4T_c \pi^2 f^2} \cdot 1^2 \cdot 1^2 = \frac{V_0^2}{4T_c \pi^2 f^2} \qquad (7\text{-}51)$$

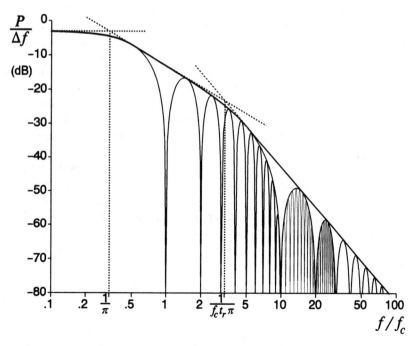

Figure 7-4   Spectral density of random digital signal, where $T_c = 10t_r$ and 0 dB represents the power $V_0^2 T_c$

and finally,

$$S(f) = \frac{V_0^2}{4T_c \pi^4 f^4 t_r^2} \sin^2(\pi f T_c) \sin^2(\pi f t_r)$$

$$\leq \frac{V_0^2}{4T_c \pi^4 f^4 t_r^2} \cdot 1^2 \cdot 1^2 = \frac{V_0^2}{4T_c \pi^4 f^4 t_r^2} \qquad (7\text{-}52)$$

These inequalities impose upper bounds on $S(f)$, which appear as dotted lines in Figure 7-4.

For frequencies such that $f\pi T_c \leq 1$, or $f \leq 1/(\pi T_c) = f_c/\pi$, Equation (7-50) is the most restrictive of the three and thus establishes the upper bound for $S(f)$. At these frequencies the power remains essentially constant. For frequencies in the interval $1/(\pi T_c) \leq f \leq 1/(\pi t_r)$, Equation (7-51) is the most restrictive. There, the power varies sinusoidally with frequency but its maximum decreases inversely with the square of the frequency. Every tenfold increase in f, called a *decade*, thus reduces the maximum power by a factor of 100, or 20 dB. For $f \geq 1/(\pi t_r)$, Equation (7-52) is the most restrictive. The power again varies sinusoidally with frequency, and its maximum now decreases inversely with the fourth power of the frequency. Every decade reduces the maximum power by a factor of 10,000, or 40 dB. These attenuation rates will be important when we study radiation from printed wiring cards containing digital logic. They are *limits;* the actual power is less at frequencies where the sine functions are not at their maximum values.

By the nature of probabilistic analyses, these results are only statistical estimates and are the results of taking an ensemble average. They are not precise for every individual case. We have assumed complete independence between successive clock periods. If we know that the data typically changes only once in every ten clock pulses, for example, then there is correlation between the data in successive clock intervals. Then the x_i variables in Equation (7-37) become correlated, and the equation is no longer true. The autocorrelation function then typically does not decrease as rapidly for $\tau \neq 0$, and the spectrum usually becomes more dense in the lower frequency ranges.

It is also possible that logic data may change at regular intervals and thus be periodic at some submultiple of the clock frequency. The spectral density then contains impulses at harmonics of these submultiples, which often are not harmonics of the clock frequency itself. An example appears in Figure 7-5, which shows a data pulse pattern that repeats itself every seven clock periods. For a periodic trapezoidal waveform, the power at the harmonic frequency f_n, obtained from Equations (7-6) and (7-1), is

$$P(f_n) = \frac{2V_0^2 t_1^2}{T_0^2} \cdot \frac{\sin^2(\pi f_n t_1)}{\pi^2 f_n^2 t_1^2} \cdot \frac{\sin^2(\pi f_n t_r)}{\pi^2 f_n^2 t_r^2} \qquad (7\text{-}53)$$

where t_r is the rise time, t_1 is the pulse duration, and T_0 is the period at which the data repeats itself, as shown in Figure 7-5. The duty cycle is t_1/T_0, which in this case is 3/7.

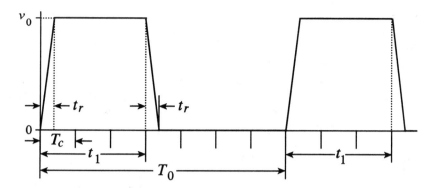

Figure 7-5   Periodic digital signal of period T_0

Due to the resemblance between Equations (7-53) and (7-49), the same observations apply to the periodic waveform as to the random process. Specifically, the power in each harmonic decreases 20 dB/decade for harmonics below the frequency $f = 1/(\pi t_r)$, and 40 dB/decade for harmonics above that frequency. The data repetition rate only affects the spacing between the harmonics, and not the envelope. Thus we can conclude that even if our statistical knowledge about the signals is sparse, the rise time gives much information about the noise spectrum of a digital logic circuit. Again it must be emphasized that using slower rise times greatly reduces high-frequency emissions from digital logic.

7-5 PROBLEMS

1. Compute the power spectral density of the random process of Equations (7-2) through (7-5).
2. A random process is defined as in Equation (7-2) but with a and f replaced by constants whose values are 1 V and 10 MHz, respectively. The random variable ϕ is defined as in Equation (7-5).
   (a) Compute the autocorrelation function of the process, using Equation (7-16).
   (b) Is this process stationary? Is it periodic?
   (c) Compute the autocorrelation function of a sample function, using Equation (7-12).
   (d) Compute the power spectral density of the process.
3. A receiver of 10-kHz bandwidth is used to detect the signal from the random process of Equations (7-2) through (7-5). Assume that the shape of the pass band is perfectly rectangular. On semi-log paper, plot the received power, as a function of the receiver center frequency f_c, for 5 MHz $< f_c <$ 15 MHz. Compare this plot with Figure 7-1.
4. Repeat Problem 3 for a receiver of 5-MHz bandwidth.
5. Repeat Problems 3 and 4 for the random process of Problem 2.
6. When the bandwidth of a receiver is halved, the received power decreases by 3 dB. Is the received noise broadband or narrowband?

7. Use the technique of Equation (7-27) to show that the power contained in the sum of two signals of *different* frequencies is always equal to the sum of the powers in the individual signals.

8. Derive Equation (7-30) using the definition of $R(\tau)$.

9. By evaluating the integral in Equation (7-42), compute the autocorrelation function of the random trapezoidal wave of Figure 7-3 with $g_r(t) = V_0 t/t_r$. Show that its Fourier transform is the power spectral density function of Equation (7-48).

10. The leading and trailing edges of the pulses of a random binary wave obey the function $g_r(t) = \frac{1}{2}V_0[1 - \cos(\pi t/t_r)]$.
    (a) Compute its power spectral density.
    (b) What is the corner frequency, in terms of t_r?
    (c) Above the corner frequency, how many decibels per decade does the power spectral density decrease?

8

RADIATION COUPLING
BETWEEN DISTANT DEVICES

Even if a circuit works correctly without interfering with itself, it still must not cause interference to other circuits or systems. Such systems may be located some distance from the noise source. For closely spaced components, we studied electric and magnetic fields separately. In this manner we were able to estimate the undesired coupling between cables and similar hardware. At greater distances, we can no longer treat electric and magnetic fields separately, so our earlier analysis is not valid. Nevertheless, we will discover that at a large distance from a noise source it is easy to combine the effects of electric and magnetic fields.

Most electric and magnetic field sources behave as combinations of dipoles. We will show that the field intensity near a static electric or magnetic dipole decreases in proportion to the inverse cube of the distance. Thus, the field becomes insignificant at moderate distances. Yet we may have observed that a time-varying electric or magnetic source, such as lightning or an arcing power transformer, may cause interference to radio receivers as far as several miles away. This occurs because, as we will also show, the electric and magnetic fields interact with each other such that their intensity decreases inversely with the distance instead of its cube. Such fields are commonly called *radiated fields*, and the resulting interference is called *radiated coupling*. We shall now discuss generation of radiated fields, show how they couple energy into circuits, and show how to minimize the radiated coupling.

8-1 SOLUTION OF MAXWELL'S EQUATIONS FOR ELECTRIC SOURCES

Every electric and magnetic field must satisfy Maxwell's equations. To find the field intensities **E** and **H** at all points in space, we must find a solution for Maxwell's equations that also satisfies the boundary conditions. Many simple solutions exist for the equations themselves, but an exact solution can be formidable for realistic boundaries. Therefore we will first examine a solution for a simple boundary condition.

A radiated electromagnetic field originates by either or both of the following two processes:

1. Alternating current flowing in a loop generates an oscillating magnetic field near the loop. As the distance from the loop increases, this magnetic field causes an electric field to develop in addition to the magnetic field.

2. AC voltage appearing on a conductor generates an oscillating electric field near the conductor. As the distance from the conductor increases, this electric field causes a magnetic field to develop in addition to the electric field.

Any voltage distribution displaces some electric charge, which can be regarded as a collection of infinitesimally small electric dipoles. Similarly, any current distribution can be regarded as a collection of small current loops. We first find a solution for the source consisting of a small electric dipole or current loop. By combining dipoles or loops on the boundaries, we can emulate any source. The solution for the entire system is then the superposition of the solutions for the individual electric dipoles and current loops. We will not actually superimpose the solutions, but they will give valuable information about the electromagnetic field.

Maxwell's equations are listed below:

$$\nabla \times \mathbf{H} = \mathbf{J} + \frac{\partial \mathbf{D}}{\partial t} \tag{8-1}$$

$$\nabla \times \mathbf{E} = -\frac{\partial \mathbf{B}}{\partial t} \tag{8-2}$$

$$\nabla \cdot \mathbf{D} = \rho \tag{8-3}$$

$$\nabla \cdot \mathbf{B} = 0 \tag{8-4}$$

Our source is an infinitesimal electric dipole consisting of two time-varying charges $\pm q(t)$, as shown in Figure 8-1. We first assume the charges to be at $(r = l/2, \theta = 0)$ and $(r = l/2, \theta = \pi)$, respectively, in a spherical coordinate system. Then we evaluate the limit as $l \to 0$ while $q(t) \cdot l$ remains finite.

The solution to Maxwell's equations gives the electric field at every point (r, ϕ, θ). We will later show that in any region where $\rho = 0$ and $\mathbf{J} = 0$, the unique solution for this source is

$$\mathbf{E} = \frac{l}{4\pi\epsilon}\left\{2\left[\frac{q(t-r/c)}{r^3} + \frac{q'(t-r/c)}{cr^2}\right]\cos\theta\,\mathbf{a}_r + \right.$$
$$\left.\left[\frac{q(t-r/c)}{r^3} + \frac{q'(t-r/c)}{cr^2} + \frac{q''(t-r/c)}{c^2r}\right]\sin\theta\,\mathbf{a}_\theta\right\} \tag{8-5}$$

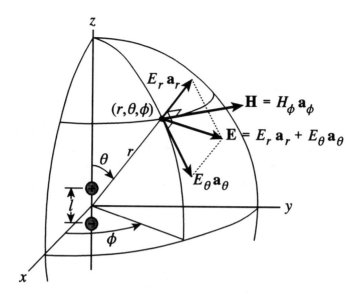

Figure 8-1   Fields of an infinitesimal oscillating electric dipole

where q' and q'' are the first and second time derivatives of $q(t)$, $\mathbf{a}_r$ and $\mathbf{a}_\theta$ are unit vectors in the respective directions, $c = 1/\sqrt{\mu\epsilon}$, and μ and ϵ are the permeability and permittivity, respectively, of the medium surrounding the dipole. For free space (the usual case), $\mu = \mu_0 = 1.26\ \mu\text{H/m}$ and $\epsilon = \epsilon_0 = 8.84\ \text{pF/m}$. The corresponding magnetic field is

$$\mathbf{H} = \frac{l}{4\pi}\left[\frac{q''(t-r/c)}{cr} + \frac{q'(t-r/c)}{r^2}\right]\sin\theta\ \mathbf{a}_\phi \qquad (8\text{-}6)$$

where $\mathbf{a}_\phi$ is a unit vector in the ϕ direction. Before proving the solution, we will study its characteristics.

Any function of $t - r/c$ is that same function of t with its time origin shifted, or *retarded*, by the time interval r/c. Thus, Equations (8-5) and (8-6) imply that at any point in space the electric and magnetic fields are proportional to a weighted sum of $q(t)$ and its first two derivatives, retarded by the time interval r/c. The physical effect is three traveling waves moving in the r direction at the velocity c, each decreasing inversely with r, r^2, or r^3. Note that neither $\mathbf{E}$ nor $\mathbf{H}$ point in the exact direction that the waves are traveling.

If $q(t)$ is a sinusoidal function, it may be written as

$$q(t) = Q\cos\omega t = \mathfrak{R}\{Qe^{j\omega t}\} \qquad (8\text{-}7)$$

Then Equations (8-5) and (8-6) become

$$\mathbf{E} = \mathfrak{R}\left\{\frac{lQe^{j\omega(t-r/c)}}{4\pi\epsilon}\left[2\left(\frac{1}{r^3}+\frac{j\omega}{cr^2}\right)\cos\theta\ \mathbf{a}_r + \left(\frac{1}{r^3}+\frac{j\omega}{cr^2}-\frac{\omega^2}{c^2r}\right)\sin\theta\ \mathbf{a}_\theta\right]\right\} \qquad (8\text{-}8)$$

and

$$\mathbf{H} = \Re\left\{\frac{lQe^{j\omega(t-r/c)}}{4\pi}\left(-\frac{\omega^2}{cr} + \frac{j\omega}{r^2}\right)\sin\theta\,\mathbf{a}_\phi\right\} \tag{8-9}$$

We next define the *phase constant* β as follows:

$$\beta = \frac{\omega}{c} = \frac{2\pi f}{c} = \frac{2\pi}{\lambda} \tag{8-10}$$

where f is the frequency and λ is the wavelength. The phase constant represents the rate of change in the phase of $\mathbf{E}$ and $\mathbf{H}$ with respect to the distance r. With this definition, Equations (8-8) and (8-9) become

$$\mathbf{E} = \Re\left\{\frac{lQe^{j(\omega t - \beta r)}}{4\pi\epsilon}\left[2\left(\frac{1}{r^3} + \frac{j\beta}{r^2}\right)\cos\theta\,\mathbf{a}_r + \left(\frac{1}{r^3} + \frac{j\beta}{r^2} - \frac{\beta^2}{r}\right)\sin\theta\,\mathbf{a}_\theta\right]\right\} \tag{8-11}$$

and

$$\mathbf{H} = \Re\left\{\frac{lQe^{j(\omega t - \beta r)}}{4\pi\sqrt{\mu\epsilon}}\left(-\frac{\beta^2}{r} + \frac{j\beta}{r^2}\right)\sin\theta\,\mathbf{a}_\phi\right\} \tag{8-12}$$

Let $\mathbf{E}_s$ and $\mathbf{H}_s$ be the phasor representations of $\mathbf{E}$ and $\mathbf{H}$, defined such that

$$\mathbf{E} = \Re\{\mathbf{E}_s e^{j\omega t}\} \tag{8-13}$$

$$\mathbf{H} = \Re\{\mathbf{H}_s e^{j\omega t}\} \tag{8-14}$$

Equations (8-11) and (8-12) are satisfied if we define these phasors to be

$$\mathbf{E}_s = \frac{lQe^{-j\beta r}}{4\pi\epsilon}\left[2\left(\frac{1}{r^3} + \frac{j\beta}{r^2}\right)\cos\theta\,\mathbf{a}_r + \left(\frac{1}{r^3} + \frac{j\beta}{r^2} - \frac{\beta^2}{r}\right)\sin\theta\,\mathbf{a}_\theta\right] \tag{8-15}$$

and

$$\mathbf{H}_s = \frac{lQe^{-j\beta r}}{4\pi\sqrt{\mu\epsilon}}\left(-\frac{\beta^2}{r} + \frac{j\beta}{r^2}\right)\sin\theta\,\mathbf{a}_\phi \tag{8-16}$$

Let $\mathbf{D}_s$ and $\mathbf{B}_s$ be the phasor equivalents of $\mathbf{D}$ and $\mathbf{B}$, defined as $\mathbf{D}_s = \epsilon\mathbf{E}_s$ and $\mathbf{B}_s = \mu\mathbf{H}_s$. Then, from the definitions of $\mathbf{D}$ and $\mathbf{B}$,

$$\mathbf{D} = \epsilon\mathbf{E} = \epsilon\Re\{\mathbf{E}_s e^{j\omega t}\} = \Re\{\epsilon\mathbf{E}_s e^{j\omega t}\} = \Re\{\mathbf{D}_s e^{j\omega t}\} \tag{8-17}$$

$$\mathbf{B} = \mu\mathbf{H} = \mu\Re\{\mathbf{H}_s e^{j\omega t}\} = \Re\{\mu\mathbf{H}_s e^{j\omega t}\} = \Re\{\mathbf{B}_s e^{j\omega t}\} \tag{8-18}$$

Similarly, let $\mathbf{J}_s = \sigma\mathbf{E}_s$ so that from Ohm's law in point form,

$$\mathbf{J} = \sigma\mathbf{E} = \sigma\Re\{\mathbf{E}_s e^{j\omega t}\} = \Re\{\sigma\mathbf{E}_s e^{j\omega t}\} = \Re\{\mathbf{J}_s e^{j\omega t}\} \tag{8-19}$$

We now substitute Equations (8-13), (8-14), (8-17), (8-18), and (8-19) into Maxwell's curl equations, Equations (8-1) and (8-2). To simplify the equations, we let Maxwell's equations apply to both the real and the imaginary parts of the phasors, thus eliminating the $\mathfrak{R}\{\dots\}$ operators. Then, since $\partial(\mathbf{D}_s e^{j\omega t})/\partial t = j\omega \mathbf{D}_s e^{j\omega t}$ and similarly for $\mathbf{B}_s$, the equations become

$$\nabla \times \mathbf{H}_s = \mathbf{J}_s + j\omega\mathbf{D}_s = (\sigma + j\omega\epsilon)\mathbf{E}_s \tag{8-20}$$

$$\nabla \times \mathbf{E}_s = -j\omega\mathbf{B}_s = -j\omega\mu\mathbf{H}_s \tag{8-21}$$

Since $\nabla \cdot (\nabla \times \mathbf{F}) \equiv 0$ for any arbitrary vector $\mathbf{F}$, it follows that for $\omega \neq 0$,

$$j\omega\nabla \cdot \mathbf{B}_s = \nabla \cdot (j\omega\mathbf{B}_s) = \nabla \cdot (\nabla \times \mathbf{E}_s) = 0$$

$$\nabla \cdot \mathbf{B}_s = 0 \tag{8-22}$$

and

$$j\omega\nabla \cdot \mathbf{D}_s + \nabla \cdot \mathbf{J}_s = \nabla \cdot (j\omega\mathbf{D}_s + \mathbf{J}_s) = \nabla \cdot (\nabla \times \mathbf{H}_s) = 0$$

$$j\omega\nabla \cdot \mathbf{D}_s = -\nabla \cdot \mathbf{J}_s \tag{8-23}$$

The continuity equation states that

$$\nabla \cdot \mathbf{J} = -\frac{\partial\rho}{\partial t} \tag{8-24}$$

In phasor form this is written as

$$\nabla \cdot \mathbf{J}_s = -j\omega\rho_s \tag{8-25}$$

where ρ_s is the phasor equivalent of the charge density ρ, defined in a manner analogous to the other phasors. Combining Equations (8-23) and (8-25) yields

$$j\omega\nabla \cdot \mathbf{D}_s = -\nabla \cdot \mathbf{J}_s = j\omega\rho_s$$

$$\nabla \cdot \mathbf{D}_s = \rho_s \tag{8-26}$$

Therefore, for $\omega \neq 0$, Maxwell's divergence equations are inherently satisfied if his curl equations are satisfied and the continuity equation applies.

We now show that Equations (8-5) and (8-6) indeed satisfy Maxwell's equations in any medium where $\rho = 0$ and $\mathbf{J} = 0$, as promised earlier. It is easier to use their phasor forms, Equations (8-15) and (8-16). We must compute $\nabla \times \mathbf{E}_s$ in spherical coordinates, and to simplify the process we express each component separately. Let $(\nabla \times \mathbf{E}_s)_\phi$ represent its ϕ component. Then

$$(\nabla \times \mathbf{E}_s)_\phi = \frac{1}{r}\left[\frac{\partial(r(E_s)_\theta)}{\partial r} - \frac{\partial(E_s)_r}{\partial\theta}\right]$$

$$= \frac{1}{r}\left[-j\beta\frac{IQe^{-j\beta r}}{4\pi\epsilon}\left(\frac{1}{r^2} + \frac{j\beta}{r} - \beta^2\right)\sin\theta + \right.$$

$$\left. \frac{IQe^{-j\beta r}}{4\pi\epsilon}\left(-\frac{2}{r^3} - \frac{j\beta}{r^2}\right)\sin\theta + \frac{IQe^{-j\beta r}}{2\pi\epsilon}\left(\frac{1}{r^3} + \frac{j\beta}{r^2}\right)\sin\theta\right] \tag{8-27}$$

$$= \frac{lQe^{-j\beta r}}{4\pi\epsilon}\left(-\frac{j\beta}{r^3} + \frac{\beta^2}{r^2} + \frac{j\beta^3}{r} - \frac{2}{r^4} - \frac{j\beta}{r^3} + \frac{2}{r^4} + \frac{2j\beta}{r^3}\right)\sin\theta$$

$$= \frac{lQe^{-j\beta r}}{4\pi\epsilon}\left(\frac{\beta^2}{r^2} + \frac{j\beta^3}{r}\right)\sin\theta = -j\mu\frac{\beta}{\sqrt{\mu\epsilon}}\frac{lQe^{-j\beta r}}{4\pi\sqrt{\mu\epsilon}}\left(\frac{j\beta}{r^2} - \frac{\beta^2}{r}\right)\sin\theta$$

$$= -j\omega\mu(H_s)_\phi$$

(8-27)
(cont'd)

Similarly, we may show that

$$(\nabla \times E_s)_r = 0 = -j\omega\mu(H_s)_r \tag{8-28}$$

and

$$(\nabla \times E_s)_\theta = 0 = -j\omega\mu(H_s)_\theta \tag{8-29}$$

Therefore, all three components of Equation (8-21) are satisfied, and thus the vector equation itself is satisfied. We may similarly show that, since $J = 0$, Equation (8-20) is also satisfied. Therefore, E_s and H_s, as given in Equations (8-15) and (8-16), satisfy Maxwell's first two equations. For $\omega \neq 0$, this implies that all four of them are satisfied. In case $\omega = 0$, it follows that $\beta = \omega/c = 0$. By computing the divergence (in spherical coordinates) of both members of Equations (8-15) and (8-16) with $\beta = 0$, we may easily show that E_s and H_s satisfy Maxwell's divergence equations if $\omega = 0$. Thus, all four equations are satisfied for every ω.

Now we must show that Equations (8-15) and (8-16) truly represent the dipole source. We enclose the dipole source with a small spherical surface $r = r_0$, concentric with the origin. This surface, together with another sphere of infinite radius, form a closed boundary for the region outside the small sphere. Therefore, if we specify the **E** and **H** fields on the boundary $r = r_0$, and also specify that **E** and **H** approach zero as $r \to \infty$, then this information uniquely determines **E** and **H** everywhere outside the small sphere. Any source inside the small sphere that causes **E** and **H** to satisfy Equations (8-15) and (8-16) for $r = r_0$ will automatically satisfy them for $r > r_0$.

For $r \ll 1/\beta = \lambda/(2\pi)$, $\beta \ll 1/r$ and all terms in Equations (8-15) and (8-16) that contain β become insignificant. Therefore, as $r \to 0$, H_s approaches zero and E_s reduces to the following expression:

$$E_s = \frac{Ql}{4\pi\epsilon r^3}(2\cos\theta\, a_r + \sin\theta\, a_\theta) \tag{8-30}$$

This is a special case of Equation (8-15), so it satisfies Maxwell's equations for stationary or time-varying fields (for all ω). The only restriction is that $r \ll \lambda/(2\pi)$. We will now show that Equation (8-30) defines the **E** field of a static electric dipole.

We first consider two static charges of strength $\pm q$, separated by the distance l. We will then take the limit as $l \to 0$ while ql remains constant. The electric scalar potential V at (r, ϕ, θ) due to the two charges is

$$V = \frac{q}{4\pi\epsilon}\left(\frac{1}{R_1} - \frac{1}{R_2}\right) = \frac{q}{4\pi\epsilon} \cdot \frac{R_2 - R_1}{R_1 R_2} = \frac{q}{4\pi\epsilon} \cdot \frac{R_2^2 - R_1^2}{R_1 R_2(R_1 + R_2)} \tag{8-31}$$

where R_1 and R_2 are the distances from (r, ϕ, θ) to the positive and negative charges, respectively, as shown in Figure 8-2. From the cosine law,

$$R_1^2 = r^2 + \left(\frac{l}{2}\right)^2 - 2r\left(\frac{l}{2}\right)\cos\theta \tag{8-32}$$

$$R_2^2 = r^2 + \left(\frac{l}{2}\right)^2 + 2r\left(\frac{l}{2}\right)\cos\theta \tag{8-33}$$

Therefore,

$$V = \frac{q}{4\pi\epsilon} \cdot \frac{[r^2 + (l/2)^2 + rl\cos\theta] - [r^2 + (l/2)^2 - rl\cos\theta]}{R_1 R_2 (R_1 + R_2)}$$

$$= \frac{q}{4\pi\epsilon} \cdot \frac{2rl\cos\theta}{R_1 R_2 (R_1 + R_2)} \tag{8-34}$$

Now we let l approach zero while ql remains constant, and the potential due to the infinitesimal dipole is

$$V = \frac{ql}{4\pi\epsilon} \cdot \frac{2r\cos\theta}{\lim\limits_{l\to 0} [R_1 R_2 (R_1 + R_2)]}$$

$$= \frac{ql}{4\pi\epsilon} \cdot \frac{2r\cos\theta}{rr(r+r)} = \frac{ql\cos\theta}{4\pi\epsilon r^2} \tag{8-35}$$

The electric field $\mathbf{E}$ due to an infinitesimal dipole is therefore

$$\mathbf{E} = -\nabla V = -\left(\frac{\partial V}{\partial r}\mathbf{a}_r + \frac{1}{r}\frac{\partial V}{\partial\theta}\mathbf{a}_\theta + \frac{1}{r\sin\theta}\frac{\partial V}{\partial\phi}\mathbf{a}_\phi\right)$$

$$= \frac{ql}{4\pi\epsilon r^3}(2\cos\theta\,\mathbf{a}_r + \sin\theta\,\mathbf{a}_\theta) \tag{8-36}$$

which is the time-domain equivalent of Equation (8-30), a solution to Maxwell's equations for $r \ll \lambda/(2\pi)$. Therefore, Equation (8-36), which gives the field of a static electric dipole, satisfies Maxwell's equations even if ql varies with time, as long as $r \ll \lambda/(2\pi)$. The frequency-domain equivalent is Equation (8-30), which specifies $\mathbf{E}_s$ for $r \ll \lambda/(2\pi)$.

Now we choose r_0, the radius of the small sphere enclosing the dipole, to be much less than $\lambda/(2\pi)$. Inside the sphere, and on its surface, $\mathbf{E}_s$ is given by Equation (8-30), and since there is no magnetic source, $\mathbf{H}_s$ is insignificant. At the surface of the sphere, these values of $\mathbf{E}_s$ and $\mathbf{H}_s$ also satisfy Equations (8-15) and (8-16) since for $r = r_0 \ll \lambda/(2\pi)$ all terms containing β are insignificant. Since the sphere forms a closed boundary for the fields outside it, Equations (8-15) and (8-16) thus specify the fields everywhere outside the sphere. Since the sphere is arbitrarily small, this includes every point except the origin.

Equations (8-15) and (8-16) thus satisfy all of Maxwell's equations and represent the source consisting of an oscillating electric dipole. They are the only func-

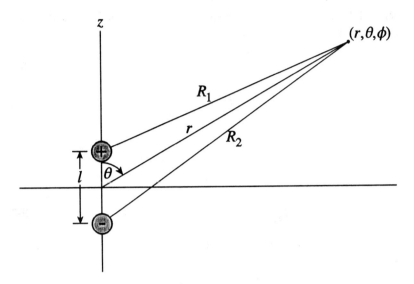

Figure 8-2 Geometry of electric dipole

tions that do so. They therefore specify the electromagnetic fields $\mathbf{E}_s$ and $\mathbf{H}_s$ everywhere in space caused by an infinitesimal oscillating electric dipole at the origin.

Any real electric-field source can be represented by a sum or integral of these infinitesimal dipoles. The resulting electromagnetic field is then the sum, or integral, of the fields due to the individual dipoles. The characteristic behavior of the fields defined by Equations (8-15) and (8-16) is therefore typical of the electromagnetic field generated by *any* electric-field source.

For $r \gg \lambda/(2\pi)$, the terms containing $1/r^2$ and $1/r^3$ in Equations (8-15) and (8-16) become insignificant. The expressions then reduce to

$$\mathbf{E}_s = \frac{lQe^{-j\beta r}}{4\pi\epsilon}\left(-\frac{\beta^2}{r}\right)\sin\theta \, \mathbf{a}_\theta = -\frac{\beta^2 lQe^{-j\beta r}}{4\pi\epsilon r}\sin\theta \, \mathbf{a}_\theta \qquad (8\text{-}37)$$

and

$$\mathbf{H}_s = \frac{lQe^{-j\beta r}}{4\pi\sqrt{\mu\epsilon}}\left(-\frac{\beta^2}{r}\right)\sin\theta \, \mathbf{a}_\phi = -\frac{\beta^2 lQe^{-j\beta r}}{4\pi\sqrt{\mu\epsilon}\,r}\sin\theta \, \mathbf{a}_\phi \qquad (8\text{-}38)$$

The region for which $r \gg \lambda/(2\pi)$ is known as the *far-field* region. In this region, $\mathbf{E}_s$ is everywhere normal to $\mathbf{H}_s$, and both fields decrease inversely with the distance r from the electric field source. The region nearer to the source, where $r \ll \lambda/(2\pi)$, is called the *near-field* region. There, the terms containing $1/r$ and $1/r^2$ in Equation (8-15) become insignificant, so $\mathbf{E}_s$ varies inversely with r^3. But Equation (8-16) contains no r^3 term, so $\mathbf{H}_s$ varies inversely with r^2. Thus, in this region, $\mathbf{E}_s$ varies more rapidly with distance than does $\mathbf{H}_s$.

In the far-field region, both $\mathbf{E}_s$ and $\mathbf{H}_s$ are perpendicular to $\mathbf{a}_r$, the direction in which the waves are traveling. We now define a new vector $\mathbf{P}_s = \mathbf{E}_s \times \mathbf{H}_s^*$, where

$\mathbf{H}_s*$ is the complex conjugate of $\mathbf{H}_s$. Then, since $\mathbf{a}_\theta \times \mathbf{a}_\phi = \mathbf{a}_r$, we note from Equations (8-37) and (8-38) that $\mathbf{P}_s$ points in the $\mathbf{a}_r$ direction, in which the waves are traveling. We will show that this is true for any sinusoidal electric field source, by adding the fields due to individual dipoles. The only restriction is that it applies only to the far-field region. In this region, the quantitative meaning of the vector $\mathbf{P}_s$ is the peak power transmitted by the electromagnetic field, but we are only interested in its direction.

The *wave impedance* is defined to be

$$Z_W = \frac{|E|_s}{|H|_s} \tag{8-39}$$

where $|E|_s$ and $|H|_s$ represent the spatial magnitudes[1] of $\mathbf{E}_s$ and $\mathbf{H}_s$. For the far-field region, where $r \gg \lambda/(2\pi)$,

$$Z_W = \frac{\dfrac{\beta^2 l Q e^{-j\beta r}}{4\pi\epsilon r}\sin\theta}{\dfrac{\beta^2 l Q e^{-j\beta r}}{4\pi\sqrt{\mu\epsilon}\,r}\sin\theta} = \frac{\sqrt{\mu\epsilon}}{\epsilon} = \sqrt{\frac{\mu}{\epsilon}} \tag{8-40}$$

Therefore, in this region, the wave impedance is constant, and it is independent of frequency and distance from the source. The wave impedance in the far-field region is called the *intrinsic impedance*, designated η. Thus,

$$\eta \doteq \sqrt{\frac{\mu}{\epsilon}} \tag{8-41}$$

In free space, the intrinsic impedance is designated η_0 and is equal to $\sqrt{\mu_0/\epsilon_0}$ or $120\,\pi\,(\approx 377)$ ohms.

In the near-field region, the wave impedance varies with the distance r from the source. It also varies with direction and frequency, and its value is generally complex. It is given by

$$Z_W = \frac{\dfrac{lQ}{4\pi\epsilon r^3}\sqrt{4\cos^2\theta + \sin^2\theta}}{\dfrac{j\beta l Q}{4\pi\sqrt{\mu\epsilon}\,r^2}\sin\theta} = \frac{1}{j\omega r\epsilon}\cdot\frac{\sqrt{3\cos^2\theta + 1}}{\sin\theta} \tag{8-42}$$

At right angles to the dipole, $\theta = \pi/2$, so $Z_W = 1/(j\omega r\epsilon)$.

---

1. We shall have occasion to refer to the *spatial* magnitude of a vector phasor as well as its *time* magnitude. The spatial and time magnitudes have meaning only for vector phasors that are expressible in the form $\mathbf{E}_s = Ee^{j\varphi}\mathbf{a}_E$, where E is a non-negative real scalar, φ is the phase angle (where $-\pi/2 < \varphi \leq \pi/2$), and $\mathbf{a}_E$ is a (real) unit vector in the direction of $\mathbf{E}_s$. To avoid confusion, we will use the notation $|E|_s$ to represent the spatial magnitude of $\mathbf{E}_s$, which is $Ee^{j\varphi}$, a complex scalar. The time magnitude of $\mathbf{E}_s$, designated $|E_s|$, is Ea_E, a real vector. The absolute magnitude of any vector phasor $\mathbf{E}_s$, designated $\|\mathbf{E}\|$, is the square root of the sum of the squares of all six components, equal to E in the above case.

8-2 SUPERPOSITION OF MULTIPLE ELECTRIC SOURCES

We will now combine the far-field effects of multiple collocated electric-dipole sources. For Equations (8-37) and (8-38) to be valid, the axis of each source must be aligned with the z axis of the coordinate system, which we have assumed for the entire analysis. If the dipoles are not parallel, which is the most general case, we must use a different coordinate system for each dipole. We do assume that the centers of all dipoles are at the same point; therefore, all coordinate systems share the same origin. Their z axes, however, point in different directions. At a given observation point, the distance r from the origin will be the same for all coordinate systems, as will the unit vector $\mathbf{a}_r$. For the same observation point, the angle θ will be different for each coordinate system. The vector $\mathbf{a}_\theta$ will always be of unit magnitude but its direction will be different in each coordinate system. The angle ϕ and the unit vector $\mathbf{a}_\phi$ are also different in each system.

Consider two electric dipoles l_1Q_1 and l_2Q_2, oscillating at the same frequency and located at a common origin. The charge quantities Q_1 and Q_2 are generally complex phasors, but we first assume that they are in phase, and we choose the time origin such that they are real. Their electric fields at a distance r, in the far field, are

$$(\mathbf{E}_1)_s = -\frac{\beta^2 l_1 Q_1 e^{-j\beta r}}{4\pi\epsilon r}\sin\theta_1\, \mathbf{a}_{\theta 1}$$

$$(\mathbf{E}_2)_s = -\frac{\beta^2 l_2 Q_2 e^{-j\beta r}}{4\pi\epsilon r}\sin\theta_2\, \mathbf{a}_{\theta 2} \qquad \textbf{(8-43)}$$

where θ_1, $\mathbf{a}_{\theta 1}$, and $\mathbf{a}_{\phi 1}$ are with respect to the coordinate system aligned with dipole 1, and similarly for θ_2, $\mathbf{a}_{\theta 2}$, and $\mathbf{a}_{\phi 2}$. The total electric field is

$$\mathbf{E}_s = (\mathbf{E}_1)_s + (\mathbf{E}_2)_s$$

$$= -\frac{\beta^2 e^{-j\beta r}}{4\pi\epsilon r}(l_1Q_1 \sin\theta_1\, \mathbf{a}_{\theta 1} + l_2Q_2 \sin\theta_2\, \mathbf{a}_{\theta 2}) \qquad \textbf{(8-44)}$$

Similarly, the total magnetic field is

$$\mathbf{H}_s = (\mathbf{H}_1)_s + (\mathbf{H}_2)_s$$

$$= -\frac{\beta^2 e^{-j\beta r}}{4\pi\epsilon\eta r}(l_1Q_1 \sin\theta_1\, \mathbf{a}_{\phi 1} + l_2Q_2 \sin\theta_2\, \mathbf{a}_{\phi 2}) \qquad \textbf{(8-45)}$$

To analyze the resulting fields, we require a common coordinate system, but neither of the above systems is convenient for our use.

The vector $\mathbf{a}_r$ is the same in both coordinate systems, and from Equations (8-44) and (8-45), all $\mathbf{E}$ and $\mathbf{H}$ vectors are perpendicular to $\mathbf{a}_r$. Therefore, all $\mathbf{E}$ and $\mathbf{H}$ vectors are in the same plane, and they appear in Figure 8-3. The angle ψ between the $\mathbf{E}$ fields (or the $\mathbf{H}$ fields) depends on the relative orientation of the dipole axes. We could calculate ψ from that information, but it is unnecessary for this analysis. From Equation (8-37),

$$\frac{|\mathbf{E}_1|_s}{|\mathbf{E}_2|_s} = \frac{-\dfrac{\beta^2 l_1 Q_1 e^{-j\beta r}}{4\pi\epsilon r}\sin\theta_1}{-\dfrac{\beta^2 l_2 Q_2 e^{-j\beta r}}{4\pi\epsilon r}\sin\theta_2} = \frac{l_1 Q_1 \sin\theta_1}{l_2 Q_2 \sin\theta_2} \tag{8-46}$$

Similarly, from Equations (8-38) and (8-41),

$$\frac{|\mathbf{H}_1|_s}{|\mathbf{H}_2|_s} = \frac{-\dfrac{\beta^2 l_1 Q_1 e^{-j\beta r}}{4\pi\epsilon\eta r}\sin\theta_1}{-\dfrac{\beta^2 l_2 Q_2 e^{-j\beta r}}{4\pi\epsilon\eta r}\sin\theta_2} = \frac{l_1 Q_1 \sin\theta_1}{l_2 Q_2 \sin\theta_2} \tag{8-47}$$

Since the above two ratios are equal, the two parallelograms in Figure 8-3 are geometrically similar. Therefore, from the geometry of the figure it is evident that $\mathbf{E}_s$ is perpendicular to $\mathbf{H}_s$. Let $\mathbf{a}_E$ be a unit vector defined as follows:

$$\mathbf{a}_E = \frac{l_1 Q_1 \sin\theta_1 \, \mathbf{a}_{\theta 1} + l_2 Q_2 \sin\theta_2 \, \mathbf{a}_{\theta 2}}{|l_1 Q_1 \sin\theta_1 \, \mathbf{a}_{\theta 1} + l_2 Q_2 \sin\theta_2 \, \mathbf{a}_{\theta 2}|}$$

$$= \frac{l_1 Q_1 \sin\theta_1 \, \mathbf{a}_{\theta 1} + l_2 Q_2 \sin\theta_2 \, \mathbf{a}_{\theta 2}}{G_E} \tag{8-48}$$

where

$$G_E = \sqrt{(l_1 Q_1 \sin\theta_1)^2 + 2 l_1 Q_1 l_2 Q_2 \sin\theta_1 \sin\theta_2 \cos\psi + (l_2 Q_2 \sin\theta_2)^2} \tag{8-49}$$

From Equation (8-44) we note that the direction of $\mathbf{E}_s$ is either the same as $\mathbf{a}_E$ or opposite to it, depending on r. Since all dipoles are in phase, Q_1 and Q_2 are real, and thus $\mathbf{a}_E$ is real, as required for a unit vector. Similarly, let $\mathbf{a}_H$ be a unit vector in the direction of $\pm\mathbf{H}_s$, defined thus:

$$\mathbf{a}_H = \frac{l_1 Q_1 \sin\theta_1 \, \mathbf{a}_{\phi 1} + l_2 Q_2 \sin\theta_2 \, \mathbf{a}_{\phi 2}}{|l_1 Q_1 \sin\theta_1 \, \mathbf{a}_{\phi 1} + l_2 Q_2 \sin\theta_2 \, \mathbf{a}_{\phi 2}|}$$

$$= \frac{l_1 Q_1 \sin\theta_1 \, \mathbf{a}_{\phi 1} + l_2 Q_2 \sin\theta_2 \, \mathbf{a}_{\phi 2}}{G_E} \tag{8-50}$$

The factor G_E describes the dipole sources and the angle of the observation point with respect to their axes, but is unrelated to the distance r. Since $\mathbf{a}_E$, $\mathbf{a}_H$, and $\mathbf{a}_r$ are mutually perpendicular unit vectors, we may conveniently define any vector in terms of these *basis* vectors. In terms of these vectors, $\mathbf{E}_s$ and $\mathbf{H}_s$ are

$$\mathbf{E}_s = -\frac{\beta^2 G_E e^{-j\beta r}}{4\pi\epsilon r}\mathbf{a}_E \tag{8-51}$$

$$\mathbf{H}_s = -\frac{\beta^2 G_E e^{-j\beta r}}{4\pi\epsilon\eta r}\mathbf{a}_H \tag{8-52}$$

where G_E is given by Equation (8-49).

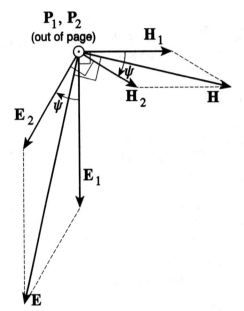

P$_1$, P$_2$
(out of page)

H$_1$

ψ

H$_2$ **H**

E$_2$ ψ

E$_1$

E

Figure 8-3 Electric and magnetic fields of two collocated non-coaxial sources

The above procedure may be extended to more than two sources in a straightforward manner, though the expressions for G_E, $\mathbf{a}_E$, and $\mathbf{a}_H$ become cumbersome. For n sources,

$$\mathbf{E}_s = \sum_{i=1}^{n} (\mathbf{E}_i)_s = -\frac{\beta^2 e^{-j\beta r}}{4\pi\epsilon r} \sum_{i=1}^{n} (l_i Q_i \sin\theta_i \, \mathbf{a}_{\theta i}) = -\frac{\beta^2 G_E e^{-j\beta r}}{4\pi\epsilon r} \, \mathbf{a}_E \qquad (8\text{-}53)$$

$$\mathbf{H}_s = \sum_{i=1}^{n} (\mathbf{H}_i)_s = -\frac{\beta^2 e^{-j\beta r}}{4\pi\epsilon\eta r} \sum_{i=1}^{n} (l_i Q_i \sin\theta_i \, \mathbf{a}_{\phi i}) = -\frac{\beta^2 G_E e^{-j\beta r}}{4\pi\epsilon\eta r} \, \mathbf{a}_H \qquad (8\text{-}54)$$

where G_E, $\mathbf{a}_E$, and $\mathbf{a}_H$ are defined as follows:

$$G_E = \left| \sum_{i=1}^{n} (l_i Q_i \sin\theta_i \, \mathbf{a}_{\theta i}) \right| \qquad (8\text{-}55)$$

$$\mathbf{a}_E = \frac{\sum_{i=1}^{n} (l_i Q_i \sin\theta_i \, \mathbf{a}_{\theta i})}{G_E} \qquad (8\text{-}56)$$

$$\mathbf{a}_H = \frac{\sum_{i=1}^{n} (l_i Q_i \sin\theta_i \, \mathbf{a}_{\phi i})}{G_E} \qquad (8\text{-}57)$$

We will never actually calculate them, since our major concern is the *nature* of the field and not its exact magnitude. For practical applications, we cannot achieve great accuracy anyhow, due to the complicated source geometry. We will use an upper bound for G_E, which we will now derive:

$$G_E = \left| \sum_{i=1}^{n} (l_i Q_i \sin\theta_i \, \mathbf{a}_{\theta i}) \right| \leq \sum_{i=1}^{n} |l_i Q_i \sin\theta_i \, \mathbf{a}_{\theta i}|$$

$$\leq \sum_{i=1}^{n} |l_i Q_i| \, |\mathbf{a}_{\theta i}| = \sum_{i=1}^{n} |l_i Q_i| \tag{8-58}$$

Physically, this upper bound disregards all directional information about the sources and assumes all of their axes to be aligned perpendicular to the direction of the observation point. Cauchy's inequality states that

$$\sum_{i=1}^{n} |l_i Q_i| \leq \sqrt{\sum_{i=1}^{n} l_i^2} \sqrt{\sum_{i=1}^{n} Q_i^2} \tag{8-59}$$

For multiple dipole sources we define l and Q to be the root-mean-squares (rms) of l_i and Q_i, respectively, for each dipole:

$$l = \sqrt{\frac{1}{n} \sum_{i=1}^{n} l_i^2} \qquad Q = \sqrt{\frac{1}{n} \sum_{i=1}^{n} Q_i^2} \tag{8-60}$$

In terms of l and Q we then have a simple and meaningful upper bound for G_E:

$$G_E \leq \sum_{i=1}^{n} |l_i Q_i| \leq \sqrt{\sum_{i=1}^{n} l_i^2} \sqrt{\sum_{i=1}^{n} Q_i^2} = n \sqrt{\frac{1}{n} \sum_{i=1}^{n} l_i^2} \sqrt{\frac{1}{n} \sum_{i=1}^{n} Q_i^2} = nlQ \tag{8-61}$$

Then, from Equations (8-51) and (8-52),

$$\|\mathbf{E}\| = \frac{\beta^2 G_E}{4\pi\epsilon r} \leq \frac{\beta^2 nlQ}{4\pi\epsilon r} \qquad \mathbf{E}_s = -\|\mathbf{E}\| e^{-j\beta r} \, \mathbf{a}_E \tag{8-62}$$

$$\|\mathbf{H}\| = \frac{\beta^2 G_E}{4\pi\epsilon\eta r} \leq \frac{\beta^2 nlQ}{4\pi\epsilon\eta r} \qquad \mathbf{H}_s = -\|\mathbf{H}\| e^{-j\beta r} \, \mathbf{a}_H \tag{8-63}$$

These closely resemble Equations (8-37) and (8-38), but without any reference to the orientations of the individual sources. Though they are inequalities, to simplify the notation we will express them as equations, with the understanding that they give upper bounds for the magnitudes of $\mathbf{E}_s$ and $\mathbf{H}_s$.

    If all dipoles are identical, then l and Q are the length and charge of any dipole. Thus, for n identical dipoles oscillating in phase, the field strength is proportional to n. To use Equations (8-62) and (8-63) for non-identical dipoles, we use the rms averages of l and Q given by Equation (8-60). If we instead use simple arithmetic averages, the error is usually negligible. Then we may regard nlQ as either the average length l times the total charge nQ, or the total length nl times the average charge Q. We will consider lQ as either of these averages, for multiple dipoles, and thus we can omit n from the field expressions. Then, for multiple sources, the expressions reduce to Equations (8-37) and (8-38) with $\theta = 90°$ and with $\mathbf{a}_\theta$ and $\mathbf{a}_\phi$ replaced by $\mathbf{a}_E$ and $\mathbf{a}_H$, respectively.

The basis vectors $\mathbf{a}_E$ and $\mathbf{a}_H$ are defined with respect to the observed field rather than the sources. When referring to a far field of a complicated or unknown electric-field source, we will normally use the basis vectors $\mathbf{a}_E$, $\mathbf{a}_H$, and $\mathbf{a}_r$. We must remember that in a far field $\mathbf{E}$ is always perpendicular to $\mathbf{H}$, and thus $\mathbf{a}_E$, $\mathbf{a}_H$, and $\mathbf{a}_r$ are perpendicular to each other. For the case analyzed here, where all dipoles are at the same point and in phase, the direction of $\mathbf{a}_E$ is sometimes called the *polarization* of the wave.

If all source dipoles are not in phase, then $(\mathbf{E}_1)_s$ and $(\mathbf{E}_2)_s$ are not in phase. The same result occurs if the distance between the sources is not small compared to $\lambda/(2\pi)$. If $(\mathbf{E}_1)_s$ and $(\mathbf{E}_2)_s$ are neither in phase nor in the same spatial direction, then their sum is not expressible in the form $Ee^{j\varphi}\mathbf{a}_E$ with E and $\mathbf{a}_E$ real. The above analysis is then not valid. To analyze this case we must convert the phasors to functions of time. We choose the time origin so that Q_1 is real, and let φ represent the phase angle between the sinusoidal functions on the dipoles. Then we may express the second dipole moment as $l_2Q_2e^{-j\varphi}$, with Q_2 real. The electric field vector phasors are

$$(\mathbf{E}_1)_s = -\frac{\beta^2 l_1 Q_1 e^{-j\beta r}}{4\pi\epsilon r}\sin\theta_1\,\mathbf{a}_{\theta 1}$$

$$(\mathbf{E}_2)_s = \frac{\beta^2 l_2 (Q_2 e^{-j\varphi})e^{-j\beta r}}{4\pi\epsilon r}\sin\theta_2\,\mathbf{a}_{\theta 2} \tag{8-64}$$

The electric fields as functions of time are

$$\mathbf{E}_1 = \mathcal{R}\{(\mathbf{E}_1)_s e^{j\omega t}\} = -\frac{\beta^2 l_1 Q_1}{4\pi\epsilon r}\cos(\omega t - \beta r)\sin\theta_1\,\mathbf{a}_{\theta 1}$$

$$\mathbf{E}_2 = \mathcal{R}\{(\mathbf{E}_2)_s e^{j\omega t}\} = -\frac{\beta^2 l_2 Q_2}{4\pi\epsilon r}\cos(\omega t - \beta r - \varphi)\sin\theta_2\,\mathbf{a}_{\theta 2} \tag{8-65}$$

The total field is

$$\mathbf{E} = \mathbf{E}_1 + \mathbf{E}_2$$

$$= -\frac{\beta^2}{4\pi\epsilon r}[l_1 Q_1\sin\theta_1\cos(\omega t - \beta r)\,\mathbf{a}_{\theta 1} + l_2 Q_2\sin\theta_2\cos(\omega t - \beta r - \varphi)\,\mathbf{a}_{\theta 2}] \tag{8-66}$$

At any fixed position with respect to the sources, this expression takes the form

$$\mathbf{E} = K_1 \cos(\omega t - \alpha_1) \, \mathbf{a}_{\theta 1} + K_2 \cos(\omega t - \alpha_2) \, \mathbf{a}_{\theta 2} \tag{8-67}$$

with K_1, K_2, α_1, and α_2 properly defined for the distance r and the angles θ_1 and θ_2. The spatial direction of the vector $\mathbf{E}$, as well as its magnitude, is constantly changing with time. A similar expression for $\mathbf{H}$ would show that its magnitude and direction are also constantly changing, though it is always perpendicular to $\mathbf{E}$. Thus in this case there would be no meaning to the unit vectors $\mathbf{a}_E$ and $\mathbf{a}_H$ discussed earlier for in-phase sources. Neither can we define a fixed direction of polarization. The tip of the vector $\mathbf{E}$ traces out a path in the plane perpendicular to r. With a suitable coordinate transformation, it may be shown[2] that this path is an ellipse. Therefore we say that this wave is *elliptically polarized*.

As stated above, an elliptically polarized wave is not representable as a single vector phasor in the form $E e^{j\varphi} \mathbf{a}_E$ with E and $\mathbf{a}_E$ real. We will now show, however, that we can represent it as two spatially perpendicular vector phasors in that form. For convenience, we choose their directions to be $\mathbf{a}_{\theta 1}$ and $\mathbf{a}_{\phi 1}$. Consider an elliptically polarized wave that results from the sum of n vector phasors $(\mathbf{E}_1)_s$ through $(\mathbf{E}_n)_s$. We first express each $(\mathbf{E}_i)_s$, for $1 \le i \le n$, in terms of $\mathbf{a}_{\theta 1}$ and $\mathbf{a}_{\phi 1}$. By vector algebra, the $\theta 1$ component of $(\mathbf{E}_i)_s$ is $(\mathbf{E}_i)_s \cdot \mathbf{a}_{\theta 1}$, and its $\phi 1$ component is $(\mathbf{E}_i)_s \cdot \mathbf{a}_{\phi 1}$. In terms of $\mathbf{a}_{\theta 1}$ and $\mathbf{a}_{\phi 1}$, $(\mathbf{E}_i)_s$ is therefore

$$(\mathbf{E}_i)_s = -\frac{\beta^2 l_i Q_i e^{-j\beta r - \varphi_i}}{4\pi\epsilon r} \sin\theta_i \, \mathbf{a}_{\theta i}$$

$$= -\frac{\beta^2 l_i Q_i e^{-j\beta r - \varphi_i}}{4\pi\epsilon r} \sin\theta_i \, [(\mathbf{a}_{\theta i} \cdot \mathbf{a}_{\theta 1})\mathbf{a}_{\theta 1} + (\mathbf{a}_{\theta i} \cdot \mathbf{a}_{\phi 1})\mathbf{a}_{\phi 1}] \tag{8-68}$$

where φ_i is the phase difference between $(\mathbf{E}_i)_s$ and $(\mathbf{E}_1)_s$. The total field $\mathbf{E}_s$ is then

$$\mathbf{E}_s = \sum_{i=1}^{n} (\mathbf{E}_i)_s = \frac{-\beta^2 e^{-j\beta r}}{4\pi\epsilon r} \sum_{i=1}^{n} \{l_i Q_i e^{-j\varphi_i} \sin\theta_i \cdot$$

$$[(\mathbf{a}_{\theta i} \cdot \mathbf{a}_{\theta 1})\mathbf{a}_{\theta 1} + (\mathbf{a}_{\theta i} \cdot \mathbf{a}_{\phi 1})\mathbf{a}_{\phi 1}]\}$$

$$= \frac{-\beta^2 e^{-j\beta r}}{4\pi\epsilon r} \sum_{i=1}^{n} [l_i Q_i e^{-j\varphi_i} \sin\theta_i \, (\mathbf{a}_{\theta i} \cdot \mathbf{a}_{\theta 1})]\mathbf{a}_{\theta 1} +$$

$$\frac{-\beta^2 e^{-j\beta r}}{4\pi\epsilon r} \sum_{i=1}^{n} [l_i Q_i e^{-j\varphi_i} \sin\theta_i \, (\mathbf{a}_{\theta i} \cdot \mathbf{a}_{\phi 1})]\mathbf{a}_{\phi 1} \tag{8-69}$$

---

2. Simon Ramo, John R. Whinnery, and Theodore Van Duzer, *Fields and Waves in Communication Electronics,* 2nd ed. (New York, N.Y.: John Wiley & Sons, Inc., 1984), pp. 276–79.

which is the sum of two perpendicular vector phasors, each in the desired form. A similar procedure is applicable to $\mathbf{H}_s$. Thus, the sum of the fields from any number of sources is always expressible as two waves of differing time phase with their $\mathbf{E}$ fields perpendicular to each other.

Since elliptically polarized waves are more difficult to analyze, we normally treat them as two spatially perpendicular components, as shown above. It will be understood that when we analyze a wave with fixed polarization, we may treat elliptically polarized waves by repeating the analysis with the $\mathbf{E}$ field rotated $90°$. Equations (8-62) and (8-63) therefore also apply to the two perpendicular components of an elliptically polarized wave, and we can thus use them for all electric-dipole sources. This technique will also be useful when we study field measurements in Chapter 12.

We assumed above that all dipole sources were in phase and added coherently. This assumption may be excessively severe, depending on the nature of the sources. If the phases of the sources are random instead of equal, then the sources add incoherently. Then, as shown in Section 7-3, the total *power* is the sum of the individual powers. Instead of Equations (8-53) and (8-54), the following relations apply:

$$\|\mathbf{E}\| = \sqrt{\sum_{i=1}^{n} \|\mathbf{E}_i\|^2} = \frac{\beta^2}{4\pi\epsilon r}\sqrt{\sum_{i=1}^{n} |l_i Q_i \sin\theta_i|^2} \le \frac{\beta^2}{4\pi\epsilon r}\sqrt{\sum_{i=1}^{n} l_i^2 |Q_i^2|} \quad \text{(8-70)}$$

$$\|\mathbf{H}\| = \sqrt{\sum_{i=1}^{n} \|\mathbf{H}_i\|^2} = \frac{\beta^2}{4\pi\epsilon\eta r}\sqrt{\sum_{i=1}^{n} |l_i Q_i \sin\theta_i|^2} \le \frac{\beta^2}{4\pi\epsilon\eta r}\sqrt{\sum_{i=1}^{n} l_i^2 |Q_i^2|} \quad \text{(8-71)}$$

As with an elliptically polarized wave, the directions of $\mathbf{E}$ and $\mathbf{H}$ are constantly changing, but here they are changing randomly. They must be perpendicular to $\mathbf{a}_r$ and to each other, but otherwise, their directions and phases are random. Therefore, in this case, it is only meaningful to speak of their absolute magnitudes $\|\mathbf{E}\|$ and $\|\mathbf{H}\|$.

From Cauchy's inequality,

$$\sum_{i=1}^{n} l_i^2 |Q_i^2| \le \sqrt{\sum_{i=1}^{n} l_i^4}\sqrt{\sum_{i=1}^{n} |Q_i^4|} \quad \text{(8-72)}$$

For multiple dipole sources adding incoherently, we define l and Q to be the following fourth-order averages:

$$l = \sqrt[4]{\frac{1}{n}\sum_{i=1}^{n} l_i^4} \qquad Q = \sqrt[4]{\frac{1}{n}\sum_{i=1}^{n} |Q_i^4|} \quad \text{(8-73)}$$

Then, from Equations (8-70) and (8-71),

$$\|\mathbf{E}\| \leq \frac{\beta^2}{4\pi\epsilon r} \sqrt[4]{\sum_{i=1}^{n} l_i^4} \sqrt[4]{\sum_{i=1}^{n} |Q_i^4|} = \frac{\beta^2}{4\pi\epsilon r} \sqrt{n} \sqrt[4]{\frac{1}{n}\sum_{i=1}^{n} l_i^4} \sqrt[4]{\frac{1}{n}\sum_{i=1}^{n} |Q_i^4|}$$

$$\leq \frac{\beta^2\sqrt{n}\, lQ}{4\pi\epsilon r} \tag{8-74}$$

$$\|\mathbf{H}\| \leq \frac{\beta^2}{4\pi\epsilon\eta r} \sqrt[4]{\sum_{i=1}^{n} l_i^4} \sqrt[4]{\sum_{i=1}^{n} |Q_i^4|} = \frac{\beta^2}{4\pi\epsilon\eta r} \sqrt{n} \sqrt[4]{\frac{1}{n}\sum_{i=1}^{n} l_i^4} \sqrt[4]{\frac{1}{n}\sum_{i=1}^{n} |Q_i^4|}$$

$$\leq \frac{\beta^2\sqrt{n}\, lQ}{4\pi\epsilon\eta r} \tag{8-75}$$

Thus, if all dipoles have the same length and charges, then the field strength is proportional to $\sqrt{n}$. This is expected, since the fields add incoherently, the total power is proportional to n, and the field strength is proportional to the square root of the power. We may substitute l and Q directly into Equations (8-74) and (8-75) if all dipole lengths and charges are nearly equal. If they are not, we replace l and Q by the fourth-order averages of Equation (8-73). Again, if we use simple arithmetic averages instead, the error is usually negligible.

At large distances from the source, as the distance r varies, the phases of $\mathbf{E}_s$ and $\mathbf{H}_s$ change much more rapidly than their amplitudes. In later chapters, we will need to consider the change in phase at distances where the change in amplitude is negligible. We now define E_0 as follows:

$$E_0 = -\|\mathbf{E}\| = -\frac{\beta^2 lQ}{4\pi\epsilon r} \tag{8-76}$$

If r changes by only a small percentage and θ remains constant, then E_0 changes only slightly and we often may treat it as constant. In terms of E_0, Equations (8-62) and (8-63) become

$$\mathbf{E}_s = E_0 e^{-j\beta r}\, \mathbf{a}_E \tag{8-77}$$

$$\mathbf{H}_s = \frac{E_0}{\eta} e^{-j\beta r}\, \mathbf{a}_H \tag{8-78}$$

With E_0 constant, Equations (8-77) and (8-78) are the equations of a *plane wave*. We may use a plane wave to approximate the fields at very large distances from a source. Equations (8-77) and (8-78) satisfy Maxwell's equations, but the source would have to be infinitely far away. As shown earlier, the plane wave is traveling in the direction of $\mathbf{P}_s = \mathbf{E}_s \times \mathbf{H}_s^*$.

8-3 SOLUTION OF MAXWELL'S EQUATIONS FOR MAGNETIC SOURCES

The preceding analysis is valid for any electric field source, since any such source can be modeled using electric dipoles. But a current loop, for example, does not directly generate an electric field at the source. We will later observe that a current loop, sometimes called a magnetic dipole, generates a magnetic field whose strength is proportional to the product of the current times the area of the loop. Consider a current $i(t)$ flowing in an infinitesimal loop of area a, as shown in Figure 8-4. This source gives a solution to Maxwell's equations that is different from that for the electric dipole. We will show that for the current-loop source the magnetic field is

$$\mathbf{H} = \frac{a}{4\pi}\left\{ 2\left[\frac{i(t-r/c)}{r^3} + \frac{i'(t-r/c)}{cr^2}\right]\cos\theta\, \mathbf{a}_r + \right.$$
$$\left. \left[\frac{i(t-r/c)}{r^3} + \frac{i'(t-r/c)}{cr^2} + \frac{i''(t-r/c)}{c^2 r}\right]\sin\theta\, \mathbf{a}_\theta\right\} \quad \textbf{(8-79)}$$

and the electric field is

$$\mathbf{E} = -\frac{\mu a}{4\pi}\left[\frac{i''(t-r/c)}{cr} + \frac{i'(t-r/c)}{r^2}\right]\sin\theta\, \mathbf{a}_\phi \quad \textbf{(8-80)}$$

For a current-loop source, the electric and magnetic fields are proportional to a weighted sum of $i(t)$ and its first two derivatives, retarded by the time interval r/c.

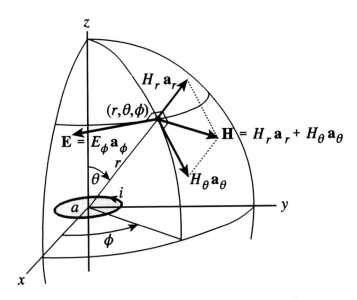

Figure 8-4 Fields of an infinitesimal oscillating current loop

As with an electric dipole source, the physical effect is three traveling waves of decreasing intensity moving in the r direction at the velocity c.

If $i(t)$ is a sinusoidal function, it may be written $i(t) = I \cos \omega t = \mathfrak{R}\{Ie^{j\omega t}\}$. A procedure similar to that used for the electric dipole shows that

$$\mathbf{H}_s = \frac{ale^{-j\beta r}}{4\pi}\left[2\left(\frac{1}{r^3} + \frac{j\beta}{r^2}\right)\cos\theta\,\mathbf{a}_r + \left(\frac{1}{r^3} + \frac{j\beta}{r^2} - \frac{\beta^2}{r}\right)\sin\theta\,\mathbf{a}_\theta\right] \tag{8-81}$$

and

$$\mathbf{E}_s = -\frac{aI\eta e^{-j\beta r}}{4\pi}\left(-\frac{\beta^2}{r} + \frac{j\beta}{r^2}\right)\sin\theta\,\mathbf{a}_\phi \tag{8-82}$$

where $\eta = \sqrt{\mu/\epsilon}$, as before.

Equations (8-81) and (8-82) satisfy Maxwell's equations, and we will show that they represent the source consisting of an oscillating current loop. They specify the electromagnetic fields $\mathbf{E}_s$ and $\mathbf{H}_s$ everywhere in space caused by an infinitesimal oscillating current loop at the origin. This solution, like the solution for an electric dipole, is unique.

Any real magnetic source can be represented by a sum or integral of these infinitesimal current loops. The resulting electromagnetic field is then the sum, or integral, of the fields due to the individual loops. The characteristic behavior of the fields defined by Equations (8-81) and (8-82) is therefore typical of the electromagnetic field generated by any magnetic field source.

For the far-field region, where $r \gg \lambda/(2\pi)$,

$$\mathbf{H}_s = \frac{ale^{-j\beta r}}{4\pi}\left(-\frac{\beta^2}{r}\right)\sin\theta\,\mathbf{a}_\theta = -\frac{\beta^2 ale^{-j\beta r}}{4\pi r}\sin\theta\,\mathbf{a}_\theta \tag{8-83}$$

$$\mathbf{E}_s = -\frac{aI\eta e^{-j\beta r}}{4\pi}\left(-\frac{\beta^2}{r}\right)\sin\theta\,\mathbf{a}_\phi = \frac{\beta^2 aI\eta e^{-j\beta r}}{4\pi r}\sin\theta\,\mathbf{a}_\phi \tag{8-84}$$

In this region, $\mathbf{E}_s$ is everywhere normal to $\mathbf{H}_s$, and both fields decrease inversely with the distance r from the magnetic field source, just as with an electric field source. Again, we note that the vector $\mathbf{P}_s$, equal to $\mathbf{E}_s \times \mathbf{H}_s{}^*$, points in the $\mathbf{a}_r$ direction, in which the wave is traveling. However, in the near-field region, the terms containing $1/r$ and $1/r^2$ in Equation (8-81) become insignificant, so $\mathbf{H}_s$ varies inversely with r^3. Equation (8-82) contains no r^3 term, so $\mathbf{E}_s$ varies inversely with r^2. Thus, in this region, $\mathbf{H}_s$ varies more rapidly with distance than does $\mathbf{E}_s$ for a magnetic field source. This is the opposite effect of an electric field source.

In the far-field region, the wave impedance of an electromagnetic field due to a current loop is

$$Z_w = \frac{|\mathbf{E}|_s}{|\mathbf{H}|_s} = \frac{\dfrac{\beta^2 aI\eta e^{-j\beta r}}{4\pi r}\sin\theta}{\dfrac{\beta^2 ale^{-j\beta r}}{4\pi r}\sin\theta} = \eta \tag{8-85}$$

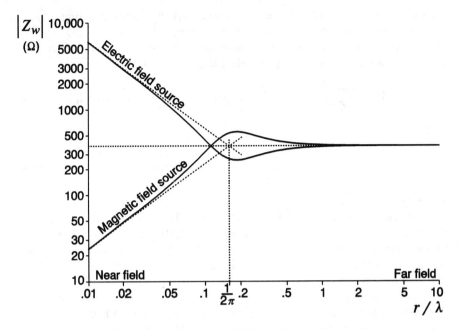

Figure 8-5   Wave impedance versus distance from a radiation source

which is the same as for an electric dipole, or 120π (≈ 377) ohms for free space. But for the near-field region, or $r \ll \lambda/(2\pi)$,

$$Z_w = \frac{\dfrac{j\beta a I \eta}{4\pi r^2} \sin\theta}{\dfrac{aI}{4\pi r^3}\sqrt{4\cos^2\theta + \sin^2\theta}} = j\omega r\mu = \frac{\sin\theta}{\sqrt{3\cos^2\theta + 1}} \qquad \textbf{(8-86)}$$

In the plane of the current loop, $\theta = \pi/2$ and thus $Z_w = j\omega r\mu$. So for small r, Z_w is large for an electric field source but small for a magnetic field source. Plots of $|Z_w|$ versus ω with $\theta = \pi/2$, for electric and magnetic sources, appear in Figure 8-5.

8-4 SUPERPOSITION OF MULTIPLE MAGNETIC SOURCES

In the far-field region, we can combine the effects of multiple collocated current-loop sources as we did with electric-dipole sources. If the planes of the current loops are not parallel, we must again use a different coordinate system for each loop. We assume that the centers of all loops are at the same point so that all coordinate systems share the same origin.

Consider two current loops $a_1 I_1$ and $a_2 I_2$, oscillating at the same frequency and located at a common origin. The currents I_1 and I_2 are complex phasors, but again we first assume them to be in phase, and we choose the time origin such that they are real. Their magnetic fields at a distance r, in the far field, are

$$(\mathbf{H}_1)_s = -\frac{\beta^2 a_1 I_1 e^{-j\beta r}}{4\pi r}\sin\theta_1\,\mathbf{a}_{\theta 1}$$

$$(\mathbf{H}_2)_s = -\frac{\beta^2 a_2 I_2 e^{-j\beta r}}{4\pi r}\sin\theta_2\,\mathbf{a}_{\theta 2} \tag{8-87}$$

where θ_1, $\mathbf{a}_{\theta 1}$, and $\mathbf{a}_{\phi 1}$ are given with respect to the coordinate system aligned with loop 1, and similarly for θ_2, $\mathbf{a}_{\theta 2}$, and $\mathbf{a}_{\phi 2}$. The total magnetic field is

$$\mathbf{H}_s = (\mathbf{H}_1)_s + (\mathbf{H}_2)_s = -\frac{\beta^2 e^{-j\beta r}}{4\pi r}(a_1 I_1\sin\theta_1\,\mathbf{a}_{\theta 1} + a_2 I_2\sin\theta_2\,\mathbf{a}_{\theta 2}) \tag{8-88}$$

Similarly, the total electric field is

$$\mathbf{E}_s = (\mathbf{E}_1)_s + (\mathbf{E}_2)_s = +\frac{\beta^2 \eta e^{-j\beta r}}{4\pi r}(a_1 I_1\sin\theta_1\,\mathbf{a}_{\phi 1} + a_2 I_2\sin\theta_2\,\mathbf{a}_{\phi 2}) \tag{8-89}$$

To analyze the resulting fields, we again require a common coordinate system.

A procedure similar to that used earlier for two electric dipoles would show that in the far field, $\mathbf{E}_s$ is always perpendicular to $\mathbf{H}_s$ for multiple current-loop sources. Let $\mathbf{a}_H$ be a unit vector in the direction of $\pm\mathbf{H}_s$, defined as follows:

$$\mathbf{a}_H = \frac{a_1 I_1\sin\theta_1\,\mathbf{a}_{\theta 1} + a_2 I_2\sin\theta_2\,\mathbf{a}_{\theta 2}}{|a_1 I_1\sin\theta_1\,\mathbf{a}_{\theta 1} + a_2 I_2\sin\theta_2\,\mathbf{a}_{\theta 2}|}$$

$$= \frac{a_1 I_1\sin\theta_1\,\mathbf{a}_{\theta 1} + a_2 I_2\sin\theta_2\,\mathbf{a}_{\theta 2}}{G_M} \tag{8-90}$$

where

$$G_M = \sqrt{(a_1 I_1\sin\theta_1)^2 + 2a_1 I_1 a_2 I_2\sin\theta_1\sin\theta_2\cos\psi + (a_2 I_2\sin\theta_2)^2} \tag{8-91}$$

Since all currents are in phase, I_1 and I_2 are real, and thus $\mathbf{a}_H$ is real, as it must be. Similarly, let $\mathbf{a}_E$ be a unit vector in the direction of $\pm\mathbf{E}_s$, defined thus:

$$\mathbf{a}_E = -\frac{a_1 I_1\sin\theta_1\,\mathbf{a}_{\phi 1} + a_2 I_2\sin\theta_2\,\mathbf{a}_{\phi 2}}{|a_1 I_1\sin\theta_1\,\mathbf{a}_{\phi 1} + a_2 I_2\sin\theta_2\,\mathbf{a}_{\phi 2}|}$$

$$= -\frac{a_1 I_1\sin\theta_1\,\mathbf{a}_{\phi 1} + a_2 I_2\sin\theta_2\,\mathbf{a}_{\phi 2}}{G_M} \tag{8-92}$$

The factor G_M describes the current-loop sources and the angle of the observation point with respect to their planes. We may define any vector in terms of the basis vectors $\mathbf{a}_E$, $\mathbf{a}_H$, and $\mathbf{a}_r$. In terms of these vectors, $\mathbf{H}_s$ and $\mathbf{E}_s$ are

$$\mathbf{H}_s = -\frac{\beta^2 G_M e^{-j\beta r}}{4\pi r}\mathbf{a}_H \tag{8-93}$$

$$\mathbf{E}_s = -\frac{\beta^2 G_M \eta e^{-j\beta r}}{4\pi r}\mathbf{a}_E \tag{8-94}$$

where G_M is given by Equation (8-91). For n current-loop sources instead of two,

$$\mathbf{H}_s = \sum_{i=1}^{n} (\mathbf{H}_i)_s = -\frac{\beta^2 e^{-j\beta r}}{4\pi r} \sum_{i=1}^{n} (a_i I_i \sin\theta_i \, \mathbf{a}_{\theta i}) = -\frac{\beta^2 G_M e^{-j\beta r}}{4\pi r} \mathbf{a}_H \qquad \textbf{(8-95)}$$

$$\mathbf{E}_s = \sum_{i=1}^{n} (\mathbf{E}_i)_s = +\frac{\beta^2 \eta e^{-j\beta r}}{4\pi r} \sum_{i=1}^{n} (a_i I_i \sin\theta_i \, \mathbf{a}_{\phi i}) = -\frac{\beta^2 \eta G_M e^{-j\beta r}}{4\pi r} \mathbf{a}_E \qquad \textbf{(8-96)}$$

where G_M, $\mathbf{a}_H$, and $\mathbf{a}_E$ are defined as follows:

$$G_M = \left| \sum_{i=1}^{n} (a_i I_i \sin\theta_i \, \mathbf{a}_{\theta i}) \right| \qquad \textbf{(8-97)}$$

$$\mathbf{a}_H = \frac{\displaystyle\sum_{i=1}^{n} (a_i I_i \sin\theta_i \, \mathbf{a}_{\theta i})}{G_M} \qquad \textbf{(8-98)}$$

$$\mathbf{a}_E = -\frac{\displaystyle\sum_{i=1}^{n} (a_i I_i \sin\theta_i \, \mathbf{a}_{\phi i})}{G_M} \qquad \textbf{(8-99)}$$

We will now derive an upper bound for G_M, following the earlier procedure used for G_E.

$$G_M = \left| \sum_{i=1}^{n} (a_i I_i \sin\theta_i \, \mathbf{a}_{\theta i}) \right| \leq \sum_{i=1}^{n} |a_i I_i \sin\theta_i \, \mathbf{a}_{\theta i}|$$

$$\leq \sum_{i=1}^{n} |a_i I_i| \, |\mathbf{a}_{\theta i}| = \sum_{i=1}^{n} |a_i I_i| \qquad \textbf{(8-100)}$$

Like the upper bound for G_E, this upper bound for G_M assumes the worst case, that all current-loop axes are aligned perpendicular to the direction of the observation point. Again, Cauchy's inequality states that

$$\sum_{i=1}^{n} |a_i I_i| \leq \sqrt{\sum_{i=1}^{n} a_i^2} \sqrt{\sum_{i=1}^{n} I_i^2} \qquad \textbf{(8-101)}$$

For multiple current loops we define a and I to be the rms of a_i and I_i, respectively, for each loop:

$$a = \sqrt{\frac{1}{n}\sum_{i=1}^{n} a_i^2} \quad I = \sqrt{\frac{1}{n}\sum_{i=1}^{n} I_i^2} \quad\quad \textbf{(8-102)}$$

In terms of a and I, the upper bound for G_M is

$$G_M \le \sum_{i=1}^{n} |a_i I_i| \le \sqrt{\sum_{i=1}^{n} a_i^2}\sqrt{\sum_{i=1}^{n} I_i^2} = n\sqrt{\frac{1}{n}\sum_{i=1}^{n} a_i^2}\sqrt{\frac{1}{n}\sum_{i=1}^{n} I_i^2} = naI \quad \textbf{(8-103)}$$

Then, from Equations (8-93) and (8-94),

$$\|\mathbf{H}\| \le \frac{\beta^2 naI}{4\pi r} \quad\quad \mathbf{H}_s = -\|\mathbf{H}\|e^{-j\beta r}\,\mathbf{a}_H \quad\quad \textbf{(8-104)}$$

$$\|\mathbf{E}\| \le \frac{\beta^2 naI\eta}{4\pi r} \quad\quad \mathbf{E}_s = -\|\mathbf{E}\|e^{-j\beta r}\,\mathbf{a}_E \quad\quad \textbf{(8-105)}$$

These inequalities resemble Equations (8-83) and (8-84), again without any reference to the orientations of the individual sources. Like Equations (8-62) and (8-63), we will treat them as equations, knowing that they give only upper bounds for $\mathbf{E}_s$ and $\mathbf{H}_s$. If all current loops have the same area and carry the same current, then a and I are the area and current of any loop. Thus, for n identical current loops oscillating in phase, the field strength is proportional to n. With dissimilar loops, we use the rms averages of a and I given by Equation (8-102), or approximate them with simple arithmetic averages. Then we may regard naI as either the average area a times the total current nI, or the total area na times the average current I. As with electric dipole sources, we will consider aI as one of these averages, for multiple loops, so we can omit n from the field expressions. They then reduce to Equations (8-83) and (8-84) with $\theta = 90°$ and with $\mathbf{a}_\theta$ and $\mathbf{a}_\phi$ replaced by $\mathbf{a}_H$ and $-\mathbf{a}_E$, respectively. We can thus use the latter equations for multiple current loops.

The above procedure allows us to refer to a far field of a complicated or unknown current-loop source by using the basis vectors $\mathbf{a}_E$, $\mathbf{a}_H$, and $\mathbf{a}_r$. For in-phase sources, the polarization is again defined to be the direction of $\mathbf{E}$. We must again remember that in a far field $\mathbf{E}$ is always perpendicular to $\mathbf{H}$, so $\mathbf{a}_E$, $\mathbf{a}_H$, and $\mathbf{a}_r$ are perpendicular to each other. If the sources are not in phase, or if their spacing is not small compared to $\lambda/(2\pi)$, the wave is elliptically polarized, as with electric-dipole sources. As shown for electric-dipole sources, the elliptically polarized wave can be resolved into two spatially perpendicular components. Equations (8-104) and (8-105) thus also apply to the two perpendicular components of an elliptically polarized wave, and we can thus use them for all current-loop sources.

The earlier assumption, that all current loops are in phase and add coherently, may be overly conservative. If we know that the currents in the various loops are unsynchronized, then their phases are random, and we may assume that the fields add incoherently. Then, following the same procedure used for electric-dipole sources, we may show that

$$\|\mathbf{H}\| \le \frac{\beta^2}{4\pi r} \sqrt[4]{\sum_{i=1}^{n} a_i^4} \sqrt[4]{\sum_{i=1}^{n} |I_i^4|} = \frac{\beta^2}{4\pi r} \sqrt{n} \sqrt[4]{\frac{1}{n}\sum_{i=1}^{n} a_i^4} \sqrt[4]{\frac{1}{n}\sum_{i=1}^{n} |I_i^4|}$$

$$\le \frac{\beta^2\sqrt{n}\, aI}{4\pi r} \tag{8-106}$$

$$\|\mathbf{E}\| \le \frac{\beta^2\eta}{4\pi r} \sqrt[4]{\sum_{i=1}^{n} a_i^4} \sqrt[4]{\sum_{i=1}^{n} |I_i^4|} = \frac{\beta^2\eta}{4\pi r} \sqrt{n} \sqrt[4]{\frac{1}{n}\sum_{i=1}^{n} a_i^4} \sqrt[4]{\frac{1}{n}\sum_{i=1}^{n} |I_i^4|}$$

$$\le \frac{\beta^2\sqrt{n}\, aI\eta}{4\pi r} \tag{8-107}$$

where a and I are defined to be the fourth-order averages

$$a = \sqrt[4]{\frac{1}{n}\sum_{i=1}^{n} a_i^4} \qquad I = \sqrt[4]{\frac{1}{n}\sum_{i=1}^{n} |I_i^4|} \tag{8-108}$$

As with incoherent electric-dipole sources, the directions and phases of $\mathbf{E}_s$ and $\mathbf{H}_s$ are changing randomly, and it is only meaningful to speak of their absolute magnitudes $\|\mathbf{E}\|$ and $\|\mathbf{H}\|$. If all loops have the same area and carry equal currents, then the field strength is proportional to $\sqrt{n}$. We may substitute a and I directly into Equations (8-106) and (8-107) if all loop areas and currents are nearly equal. If not, we replace a and I by the fourth-order averages of Equation (8-108), or approximate them with simple arithmetic averages.

Due to the similarity between Equations (8-104) and (8-63), and between Equations (8-105) and (8-62), the nature of a far field is the same regardless of whether the source is electric dipoles or current loops. The directions of $\mathbf{a}_E$, $\mathbf{a}_H$, and $\mathbf{a}_r$ depend only on the observed field and are the same for both types of source. If we do not know the type of source for a far field, we may consider the source as either type and choose IQ or aI accordingly.

As with an electric-dipole source, we may approximate the fields at large distances from a current-loop source by a plane wave. Here we define E_0 thus:

$$E_0 = -\frac{\beta^2 aI\eta}{4\pi r} \tag{8-109}$$

In terms of E_0, Equations (8-104) and (8-105) become

$$\mathbf{H}_s = \frac{E_0}{\eta} e^{-j\beta r}\, \mathbf{a}_H \tag{8-110}$$

$$\mathbf{E}_s = E_0 e^{-j\beta r}\, \mathbf{a}_E \tag{8-111}$$

The plane wave is again traveling in the direction of $\mathbf{E}_s \times \mathbf{H}_s^*$. Since Equations (8-111) and (8-110) are respectively identical to Equations (8-77) and (8-78), it makes no difference whether the source of a plane wave is electric, magnetic, or a combination thereof.

8-5 MAGNETIC FIELD SOURCE CHARACTERISTICS

A high-frequency magnetic field source is a loop of high-frequency current. Often, the loop consists of two parallel conductors connected to components at their ends. Because the current flows through one conductor and returns in the other, it is differential-mode current. To analyze fields due to this current, we find the field caused by a rectangular loop of current i, as shown in Figure 8-6. We first consider only the portion of the field created by the two current segments flowing in the $\pm x$ direction.

To simplify the analysis we use the magnetic vector potential, previously used in Chapter 3. The magnetic vector potential **A** due to a current i flowing through a short conductor of length l is[3,4]

$$\mathbf{A} = \frac{\mu i l}{4\pi R}\mathbf{a}_l \tag{8-112}$$

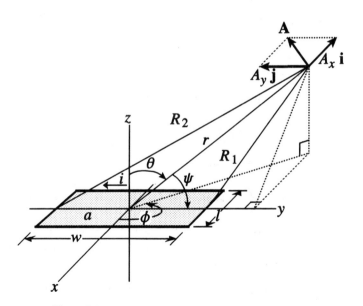

Figure 8-6   Magnetic vector potential due to a current loop

3. William H. Hayt, *Engineering Electromagnetics,* 5th ed. (New York, N.Y.: McGraw-Hill Publishing Company, 1989), pp. 249–57.

4. John D. Kraus, *Electromagnetics,* 4th ed. (New York, N.Y.: McGraw-Hill Publishing Company, 1992), pp. 258–63.

The magnetic vector potential of a current element behaves much like the electric scalar potential V of an electric charge. The direction of $\mathbf{A}$ is the direction of the current that causes it and does not depend on the direction of the observation point from the source. Let A_x be the x component of the total vector potential $\mathbf{A}$. The two current segments flowing in the x direction determine A_x, which is

$$A_x = \frac{\mu il}{4\pi}\left(\frac{1}{R_2} - \frac{1}{R_1}\right) = -\frac{\mu il}{4\pi}\left(\frac{R_2 - R_1}{R_1 R_2}\right) \tag{8-113}$$

By the same reasoning as used in Equations (8-32) through (8-34), we may show that

$$A_x = -\frac{\mu il}{4\pi} \cdot \frac{2rw\cos\psi}{R_1 R_2 (R_1 + R_2)} \tag{8-114}$$

As shown for an electric scalar potential due to a dipole of charge, in Equation (8-35), the vector potential component due to the two current segments is, for $r \gg l_i$,

$$A_x \approx -\frac{\mu ilw\cos\psi}{4\pi r^2} = -\frac{\mu ai\cos\psi}{4\pi r^2} \tag{8-115}$$

where a is the area of the loop. The projection of r onto the xy plane is $r\sin\theta$, and the projection of $(r\sin\theta)$ onto the y axis is $(r\sin\theta)\sin\phi$. This projection is also the projection of r onto the y axis, which is $r\cos\Psi$. Therefore, $\cos\Psi = \sin\theta\sin\phi$, and

$$A_x = -\frac{\mu ai}{4\pi r^2}\sin\theta\sin\phi \tag{8-116}$$

We may similarly show that A_y, the y component of the total vector potential $\mathbf{A}$, is

$$A_y = +\frac{\mu ia}{4\pi r^2}\sin\theta\cos\phi \tag{8-117}$$

The total vector potential $\mathbf{A}$ is therefore

$$\mathbf{A} = A_x\mathbf{i} + A_y\mathbf{j} = \frac{\mu ia}{4\pi r^2}\sin\theta(-\sin\phi\,\mathbf{i} + \cos\phi\,\mathbf{j}) = \frac{\mu ia}{4\pi r^2}\sin\theta\,\mathbf{a}_\phi \tag{8-118}$$

where $\mathbf{i}$ and $\mathbf{j}$ are unit vectors in the x and y directions, respectively. It is shown in the references that $\mathbf{B} = \nabla \times \mathbf{A}$. From the vector potential we obtain the magnetic field intensity of a rectangular current loop:

$$\begin{aligned}
\mathbf{H} &= \frac{\mathbf{B}}{\mu} = \frac{\nabla \times \mathbf{A}}{\mu} = \frac{1}{r\sin\theta}\left[\frac{\partial(A_\phi\sin\theta)}{\partial\theta}\right]\mathbf{a}_r + \frac{1}{r}\left[-\frac{\partial(rA_\phi)}{\partial r}\right]\mathbf{a}_\theta \\
&= \frac{ia}{4\pi}\left[\frac{1}{r^3\sin\theta}(2\sin\theta\cos\theta)\mathbf{a}_r - \frac{\sin\theta}{r}\left(-\frac{1}{r^2}\right)\mathbf{a}_\theta\right] \\
&= \frac{ia}{4\pi r^3}(2\cos\theta\,\mathbf{a}_r + \sin\theta\,\mathbf{a}_\theta)
\end{aligned} \tag{8-119}$$

A planar current loop of any shape may be approximated by a multitude of rectangular current loops, as shown in Figure 8-7. The fictitious currents flowing across the loop total zero, and the only actual current is in the conductor outlining the loop. The total field intensity is the sum of the fields due to the individual loops, and the total area of the loop is the sum of the individual areas. If $\mathbf{H}_k$ is the field due to loop k of area a_k, n is the number of rectangular loops, $\mathbf{H}$ is the total magnetic field intensity, and a is the total loop area, then

$$\mathbf{H} = \sum_{k=1}^{n} \mathbf{H}_k = \sum_{k=1}^{n} \left[\frac{ia_k}{4\pi r^3} (2\cos\theta\, \mathbf{a}_r + \sin\theta\, \mathbf{a}_\theta) \right]$$

$$= \frac{i\left(\sum_{k=1}^{n} a_k \right)}{4\pi r^3} (2\cos\theta\, \mathbf{a}_r + \sin\theta\, \mathbf{a}_\theta)$$

$$= \frac{ia}{4\pi r^3} (2\cos\theta\, \mathbf{a}_r + \sin\theta\, \mathbf{a}_\theta) \qquad \text{(8-120)}$$

Equation (8-119) is thus valid for any planar loop, of area a, lying in the xy plane.

We will now show that Equations (8-81) and (8-82) represent the source consisting of an oscillating current loop, as promised earlier. We use the same procedure as for the electric dipole source. Equation (8-81), with $r \ll \lambda/(2\pi) = 1/\beta$, is approximately

$$\mathbf{H}_s = \frac{aI}{4\pi r^3} (2\cos\theta\, \mathbf{a}_r + \sin\theta\, \mathbf{a}_\theta) \qquad \text{(8-121)}$$

which is the phasor equivalent of Equation (8-119) and satisfies Maxwell's equations for $r \ll \lambda/(2\pi)$. Therefore, for $r \ll \lambda/(2\pi)$, the $\mathbf{H}$ field due to a current $i(t)$

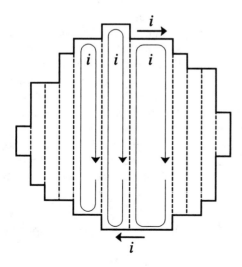

Figure 8-7 Summation of rectangular current loops onto loop of arbitrary shape

flowing in a small loop of area a is given by Equation (8-119), even if i is not constant. If we enclose the current-loop source by the small sphere $r = r_0$, where $r_0 \ll \lambda/(2\pi)$, then, within the sphere, $\mathbf{E}$ is negligible and $\mathbf{H}$ is given by Equation (8-119). The fields on the surface of the sphere thus agree with Equations (8-81) and (8-82). The sphere, together with the point at infinity, form a closed boundary surface, as explained earlier in the analysis of the electric dipole. Since the fields on the surface $r = r_0$ are the same as given by Equations (8-81) and (8-82), and since $\mathbf{H} = \mathbf{E} = 0$ at infinity, the fields everywhere outside the sphere must be those given by Equations (8-81) and (8-82). Thus we have shown that Equations (8-81) and (8-82) indeed represent a source consisting of a current I in a loop of area a. This is true for any current loop whose largest diameter is small compared to $\lambda/(2\pi)$. For larger loops, the currents are not equal everywhere on the loop, and the above equations are not usable.

The electric far field due to the current loop, for $r \gg \lambda/(2\pi)$, is given by Equation (8-84), or Equation (8-105) for multiple current loops. In free space, the time magnitude of $\mathbf{E}_s$ from Equation (8-105) is

$$|\mathbf{E}_s| = \frac{\beta^2 I a \eta_0}{4\pi r}\, \mathbf{a}_E$$

$$= \frac{(2\pi f\sqrt{\mu_0\epsilon_0})^2 I a \sqrt{\mu_0/\epsilon_0}}{4\pi r}\, \mathbf{a}_E$$

$$= \frac{\pi\mu_0\sqrt{\mu_0\epsilon_0}\, f^2 I a}{r}\, \mathbf{a}_E$$

$$= \frac{1.316 f^2 I a}{r}\, \mathbf{a}_E \;\; \mu\text{V/m} \qquad \textbf{(8-122)}$$

with f expressed in MHz, I in amperes, a in cm^2, and r in meters.

If the current loop sources are near a flat conducting surface, the reflection from the surface will add to the incident wave. The tangential electric field intensity must be zero at the conducting plane. To satisfy Maxwell's equations with this restriction, the field intensity everywhere must be just as though there were an image of each current loop behind the plane, as shown in Figure 8-8. The fictitious current in each image must be equal to the current in the actual loop, and must flow in the opposite direction. If $(\mathbf{E}_1)_s$ is the field due to the current loops and $(\mathbf{E}_2)_s$ is the field due to their images, the total field $\mathbf{E}_s$ is the sum of these. Let $\mathbf{R}_1$ be a vector to the point (r, θ, ϕ) from the center of the source loops, and let $\mathbf{R}_2$ be a vector from the center of their images. For $r \gg \lambda/(2\pi)$, the far field, the electric field intensity $\mathbf{E}_s$ at (r, θ, ϕ) is, from Equation (8-105),

$$\mathbf{E}_s = (\mathbf{E}_1)_s + (\mathbf{E}_2)_s = -\frac{\beta^2 a I \eta}{4\pi R_1}(e^{-j\beta R_1}\, \mathbf{a}_{E1} + e^{-j\beta R_2}\, \mathbf{a}_{E2}) \qquad \textbf{(8-123)}$$

where $R_1 = |\mathbf{R}_1|$, $R_2 = |\mathbf{R}_2|$, and $\mathbf{a}_{E1}$ and $\mathbf{a}_{E2}$ are the directions of $(\mathbf{E}_1)_s$ and $(\mathbf{E}_2)_s$, respectively. We note that $\mathbf{R}_1$ and $\mathbf{R}_2$ project from the source and its image at different angles with respect to the source axes. Since the directional characteristics of the sources are unknown, $(\mathbf{E}_1)_s$ and $(\mathbf{E}_2)_s$ might be in any directions perpendicular to $\mathbf{R}_1$ and $\mathbf{R}_2$, respectively. Since, in general, they are not in phase, the resulting wave may be elliptically polarized. To analyze it, we must consider its two perpendicular components separately. If h is the distance between the loop and the conducting plane, then for $r \gg h$, $R_1 \approx R_2 \approx r$, so

$$\mathbf{E}_s = -\frac{\beta^2 I a \eta}{4\pi r}(e^{-j\beta R_1}\, \mathbf{a}_{E1} + e^{-j\beta R_2}\, \mathbf{a}_{E2})$$

$$= -\frac{\beta^2 I a \eta}{4\pi r}\exp\left(-j\beta\,\frac{R_1+R_2}{2}\right)\left[\exp\left(-j\beta\,\frac{R_1-R_2}{2}\right)\mathbf{a}_{E1} + \exp\left(-j\beta\,\frac{-R_1+R_2}{2}\right)\mathbf{a}_{E2}\right]$$

$$(8\text{-}124)$$

By the cosine law, since $R_1 \approx R_2 \approx r$,

$$R_2 - R_1 = \frac{R_2^2 - R_1^2}{R_2 + R_1} = \frac{[r^2 + h^2 + 2rh\cos\theta] - [r^2 + h^2 - 2rh\cos\theta]}{R_2 + R_1}$$

$$= \frac{4rh\cos\theta}{R_2 + R_1} \approx 2h\cos\theta \qquad (8\text{-}125)$$

and thus,

$$\mathbf{E}_s = -\frac{\beta^2 I a \eta e^{-j\beta r}}{4\pi r}(e^{j\beta h\cos\theta}\, \mathbf{a}_{E1} + e^{-j\beta h\cos\theta}\, \mathbf{a}_{E2}) \qquad (8\text{-}126)$$

We will choose the two perpendicular components to be in the directions of $\mathbf{a}_{E1}$ and $\mathbf{a}_{H1}$. From vector algebra, the $\mathbf{a}_{E1}$ component of $\mathbf{E}_s$ is $\mathbf{E}_s \cdot \mathbf{a}_{E1}$, and its $\mathbf{a}_{H1}$ component is $\mathbf{E}_s \cdot \mathbf{a}_{H1}$. Using these basis vectors we may express $\mathbf{E}_s$ as follows:

$$\mathbf{E}_s = \frac{-\beta^2 I a \eta e^{-j\beta r}}{4\pi r}(e^{j\beta h\cos\theta} + e^{-j\beta h\cos\theta}\,(\mathbf{a}_{E2}\cdot \mathbf{a}_{E1})]\,\mathbf{a}_{E1} +$$

$$= \frac{-\beta^2 I a \eta e^{-j\beta r}}{4\pi r}(e^{-j\beta h\cos\theta}\,(\mathbf{a}_{E2}\cdot \mathbf{a}_{H1})\,\mathbf{a}_{H1} \qquad (8\text{-}127)$$

The two exponentials in the brackets in Equation (8-127) are phasors whose magnitudes and phases differ, in general. The time magnitude of the $\mathbf{a}_{E1}$ component of $\mathbf{E}_s$ is greatest, however, when the two phasors are in phase, if $\mathbf{a}_{E2}\cdot \mathbf{a}_{E1} > 0$. They are in phase if $h = n\pi/(\beta|\cos\theta|)$ for n an integer, since both exponents then become multiples of $j\pi$, and

$$\mathbf{E}_s = \frac{-\beta^2 I a \eta e^{-j\beta r}}{4\pi r}[e^{jn\pi} + e^{-jn\pi}\,(\mathbf{a}_{E2}\cdot \mathbf{a}_{E1})]\,\mathbf{a}_{E1} +$$

$$\frac{-\beta^2 I a \eta e^{-j\beta r}}{4\pi r}e^{-jn\pi}\,(\mathbf{a}_{E2}\cdot \mathbf{a}_{H1})]\,\mathbf{a}_{H1} \qquad (8\text{-}128)$$

$$= \frac{-\beta^2 I a \eta e^{-j\beta r}}{4\pi r} [(\pm 1) + (\pm 1)(\mathbf{a}_{E2} \cdot \mathbf{a}_{E1})] \mathbf{a}_{E1} +$$

$$\frac{-\beta^2 I a \eta e^{-j\beta r}}{4\pi r}(\pm 1)(\mathbf{a}_{E2} \cdot \mathbf{a}_{H1}) \mathbf{a}_{H1} \qquad \begin{array}{l}\textbf{(8-128)} \\ \textbf{(cont.)}\end{array}$$

The length of the reflected path then differs from that of the incident path by a multiple of a wavelength, and the two signals then add in phase. If $(\mathbf{E}_1)_s$ and $(\mathbf{E}_2)_s$ are in the same spatial direction, $\mathbf{a}_{E2} \cdot \mathbf{a}_{E1} = 1$, and

$$\mathbf{E}_s = \pm \frac{-\beta^2 I a \eta e^{-j\beta r}}{4\pi r}(1+1)\mathbf{a}_{E1} \pm \frac{-\beta^2 I a \eta e^{-j\beta r}}{4\pi r}(0)\mathbf{a}_{H1}$$

$$= \mp \frac{\beta^2 I a \eta e^{-j\beta r}}{2\pi r}\mathbf{a}_{E1} \qquad \textbf{(8-129)}$$

Their sum is therefore double the original field strength, which is the worst case.

If $\mathbf{a}_{E2} \cdot \mathbf{a}_{E1} < 0$, the maximum field strength occurs when the exponentials in the brackets in Equation (8-127) are 180° out of phase. This occurs for $h = (2n + 1)\pi/(2\beta|\cos\theta|)$, when the length of the reflected path differs from that of the incident path by an odd number of half wavelengths. In either case, Equation (8-129) gives the maximum (worst-case) field strength.

Since the directions of $(\mathbf{E}_1)_s$ and $(\mathbf{E}_2)_s$ are nearly impossible to predict, we assume $\mathbf{E}_s$ to be the worst-case value given by Equation (8-129). This gives a

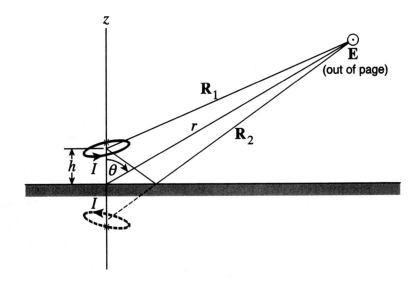

Figure 8-8 Current loop over conductive plane

least upper bound for $|\mathbf{E}_s|$, the time magnitude of $\mathbf{E}_s$. In free space, this bound is therefore

$$
\begin{aligned}
|\mathbf{E}_s| &= \frac{\beta^2 I a \eta_0}{2\pi r}\, \mathbf{a}_E \\[2mm]
&= \frac{(2\pi f\sqrt{\mu_0\epsilon_0})^2 I a\sqrt{\mu_0/\epsilon_0}}{2\pi r}\, \mathbf{a}_E \\[2mm]
&= \frac{2\pi\mu_0\sqrt{\mu_0\epsilon_0}\, f^2 I a}{r}\, \mathbf{a}_E \\[2mm]
&= \frac{2.632 f^2 I a}{r}\, \mathbf{a}_E \; \mu\text{V/m}
\end{aligned}
\qquad (8\text{-}130)
$$

with f expressed in megahertz, I in amperes, a in square centimeters, and r in meters, as before. To eliminate the effects caused by multiple reflections, EMI measurements are normally made over a conductive floor, as we will discuss in Chapter 12. To predict the field consistent with these measurements, we must use Equation (8-130).

By solving for a, we may calculate the maximum allowable loop area for a given absolute field strength magnitude $\|\mathbf{E}\|$ at a given distance from the source:

$$
a = \frac{\|\mathbf{E}\| r}{2.632 f^2 I}\; \text{cm}^2 = \frac{0.380\,\|\mathbf{E}\| r}{f^2 I}\; \text{cm}^2
\qquad (8\text{-}131)
$$

with $\|\mathbf{E}\|$ expressed in microvolts per meter, f in megahertz, I in amperes, and r in meters. For example, if $\|\mathbf{E}\|$ must be limited to 100 μV/m at $r = 3$ m, and if $I = 25$ mA at $f = 30$ MHz, the maximum allowable area a is

$$
a = \frac{0.380 \cdot 100 \cdot 3}{30^2 \cdot 0.025}\; \text{cm}^2 = 5.07\; \text{cm}^2
\qquad (8\text{-}132)
$$

The area a becomes smallest if the return path for the current is as near as possible to the incident path. With a stripline, discussed in Chapter 6, the distance between the incident and return current paths is only the thickness of its dielectric. This minimizes the area a, for a given lead length. Thus, a stripline reduces radiated coupling as well as the capacitive and inductive coupling discussed in Chapter 3. This observation reaffirms the desirability of a stripline or a solid ground plane, which behaves similarly.

In Chapter 7, we studied the frequency spectra of typical noise sources. We investigated, in particular, digital logic using trapezoidal pulses with rise times designated t_r. For such circuits, the power decreases 20 dB/decade up to the frequency $f = 1/\pi t_r$, and 40 dB/decade above that frequency. This power could represent either a voltage or a current source. For the present analysis it will represent the loop current, and its units will be amperes^2. The current I is then equal to the square root of the power. Due to the definition of the decibel, it decreases the same number of decibels per decade. For a random binary sequence of pulses controlled by a clock of frequency f_c and period T_c, the noise current is broadband and is given by

$$|I(f, \Delta f)| = \sqrt{P(f, \Delta f)} = \sqrt{2S(f)\,\Delta f}$$

$$= I_0 \sqrt{\frac{T_c \Delta f}{2}} \cdot \frac{\sin(\pi f T_c)}{\pi f T_c} \cdot \frac{\sin(\pi f t_r)}{\pi f t_r} \tag{8-133}$$

where I_0 is the peak loop current. To obtain the field intensity as a function of frequency, we substitute the current I from Equation (8-133) into Equation (8-130). The resulting field intensity is

$$|\mathbf{E}_s(f, \Delta f)| = \frac{2I_0 a \mu_0 \pi f^2}{rc} \sqrt{\frac{T_c \Delta f}{2}} \left| \frac{\sin(\pi f T_c)}{\pi f T_c} \cdot \frac{\sin(\pi f t_r)}{\pi f t_r} \right| \mathbf{a}_E$$

$$= \frac{\sqrt{2} I_0 a \mu_0}{\pi r c t_r} \sqrt{\frac{\Delta f}{T_c}} \, |\sin(\pi f T_c)\sin(\pi f t_r)| \, \mathbf{a}_E$$

$$= \left(\frac{\sqrt{2}\,\mu_0}{\pi c} \right) \frac{I_0 a}{r t_r} \sqrt{\frac{\Delta f}{T_c}} \, |\sin(\pi f T_c)\sin(\pi f t_r)| \, \mathbf{a}_E \tag{8-134}$$

Let K_{loop} represent the parenthesized constant in Equation (8-134), or

$$K_{loop} = \frac{\sqrt{2}\,\mu_0}{\pi c} = 1.88 \cdot 10^{-15} \frac{\text{H s}}{\text{m}^2} = 1.88 \cdot 10^{-15} \frac{\text{V s}^2}{\text{A m}^2}$$

$$= 0.188 \frac{\mu\text{V } \mu\text{s}^2}{\text{A cm}^2} \tag{8-135}$$

Then, with I_0 expressed in amperes, r in meters, a in square centimeters, f and Δf in megahertz, and T_c and t_r in microseconds, Equation (8-134) becomes

$$|\mathbf{E}_s(f, \Delta f)| = K_{loop} \frac{I_0 a}{r t_r} \sqrt{\frac{\Delta f}{T_c}} \, |\sin(\pi f T_c)\sin(\pi f t_r)| \, \mathbf{a}_E$$

$$= 0.188 \cdot \frac{I_0 a}{r t_r} \sqrt{\frac{\Delta f}{T_c}} \, |\sin(\pi f T_c)\sin(\pi f t_r)| \, \mathbf{a}_E \; \mu\text{V/m} \tag{8-136}$$

Since the magnitudes of the sine functions cannot exceed 1,

$$|\mathbf{E}_s(f, \Delta f)| \le K_{loop} \frac{I_0 a}{r t_r} \sqrt{\frac{\Delta f}{T_c}} \, \mathbf{a}_E \tag{8-137}$$

Furthermore, since $\sin(\pi f t_r) \leq \pi f t_r$,

$$|\mathbf{E}_s(f, \Delta f)| \leq K_{loop} \frac{I_0 a}{r t_r} \sqrt{\frac{\Delta f}{T_c}} \, \pi f t_r \, \mathbf{a}_E \qquad \text{(8-138)}$$

and, since $\sin(\pi f T_c) \leq \pi f T_c$,

$$|\mathbf{E}_s(f, \Delta f)| \leq K_{loop} \frac{I_0 a}{r t_r} \sqrt{\frac{\Delta f}{T_c}} (\pi f T_c)(\pi f t_r) \, \mathbf{a}_E \qquad \text{(8-139)}$$

All three inequalities, Equations (8-137), (8-138), and (8-139), must be satisfied. For $\pi f T_c \leq 1$, or $f \leq 1/(\pi T_c)$, Equation (8-139) imposes a lesser upper bound and is the inequality that must be satisfied. Therefore, the upper bound of $|\mathbf{E}_s|$ is proportional to f^2 and thus increases 40 dB per decade until $f = 1/(\pi T_c)$. For frequencies in the interval $1/(\pi T_c) \leq f \leq 1/(\pi t_r)$, Equation (8-138) dominates. There, $|\mathbf{E}_s|$ varies sinusoidally with frequency, but its upper bound is proportional to f and thus increases 20 dB per decade. For $f \geq 1/(\pi t_r)$, Equation (8-137) dominates and, though $|\mathbf{E}_s|$ still varies sinusoidally with frequency, its upper bound is constant. Thus, in this interval the upper bound is simple to calculate. From Equations (8-135) and (8-137), if $I_0 = 1$ A, $a = 1$ cm^2, $r = 1$ m, $T_c = 1$ μs, and $t_r = 0.1$ μs,

$$\frac{|\mathbf{E}_s(f, \Delta f)|}{\sqrt{\Delta f}} \leq \frac{K_{loop} I_0 a}{r t_r \sqrt{T_c}} \, \mathbf{a}_E$$

$$= \frac{0.188 \cdot 1 \cdot 1}{1 \cdot 0.1 \cdot \sqrt{1}} \frac{\mu V}{m\sqrt{MHz}} \, \mathbf{a}_E = 1.88 \frac{\mu V}{m\sqrt{MHz}} \, \mathbf{a}_E \qquad \text{(8-140)}$$

To express this field strength in dB$_{\mu V/m/MHz}$, we use the definition from Equation (7-23):

$$E_{dB\mu V/m/MHz} = 10 \log \frac{|\mathbf{E}_s(f, \Delta f)|^2/\Delta f}{1 \, \mu V^2/m^2/MHz}$$

$$\leq 10 \log (1.88^2) = 5.48 \text{ dB}_{\mu V/m/MHz} \qquad \text{(8-141)}$$

The plot of $|\mathbf{E}_s|/\sqrt{\Delta f}$, expressed in dB$_{\mu V/m/MHz}$ for the above values of I_0, a, r, T_c, and t_r, appears in Figure 8-9. As noted in Chapter 7, the envelope of the function is more important than the individual nulls.

If the current wave is periodic instead of random, then the noise is narrowband. We must use Equation (7-53) with V_0 replaced by I_0, instead of Equation (8-133). At the nth harmonic of the frequency of repetition, the current is

$$I(f) = \frac{\sqrt{2} I_0 t_1}{T_0} \cdot \frac{\sin(\pi f_n t_1)}{\pi f_n t_1} \cdot \frac{\sin(\pi f_n t_r)}{\pi f_n t_r} \qquad \text{(8-142)}$$

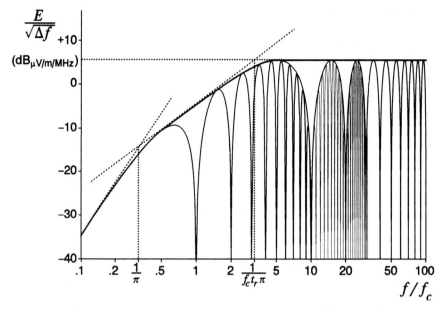

Figure 8-9 Spectral density of radiation 1 meter away from a 1 cm^2 loop carrying 1 ampere, with $T_c = 1$ μs and $t_r = 0.1$ μs

where t_1 is the pulse duration, and T_0 is the period of repetition. The field intensity is then

$$|\mathbf{E}_s(f_n)| = \frac{I_0 a \mu_0 \omega_n^2}{2\pi rc} \cdot \frac{\sqrt{2}\,t_1}{T_0} \left| \frac{\sin(\pi f_n t_1)}{\pi f_n t_1} \cdot \frac{\sin(\pi f_n t_r)}{\pi f_n t_r} \right| \mathbf{a}_E$$

$$= \frac{2\sqrt{2}\,I_0 a \mu_0}{\pi rc t_r T_0} |\sin(\pi f_n t_1)\sin(\pi f_n t_r)|\,\mathbf{a}_E$$

$$= \frac{2\sqrt{2}\,\mu_0}{\pi c} \cdot \frac{I_0 a}{rt_r T_0} |\sin(\pi f_n t_1)\sin(\pi f_n t_r)|\,\mathbf{a}_E$$

$$= 2K_{loop} \cdot \frac{I_0 a}{rt_r T_0} |\sin(\pi f_n t_1)\sin(\pi f_n t_r)|\,\mathbf{a}_E \qquad \textbf{(8-143)}$$

where K_{loop} is defined by Equation (8-135). With I_0 expressed in amperes, r in meters, a in square centimeters, f in megahertz, and T_0, t_1, and t_r in microseconds, Equation (8-143) becomes

$$|\mathbf{E}_s(f_n)| = \frac{0.377\,I_0 a}{rt_r T_0} |\sin(\pi f_n t_1)\sin(\pi f_n t_r)|\,\mathbf{a}_E \ \mu\text{V/m} \qquad \textbf{(8-144)}$$

The spectrum will contain discrete frequencies, but their envelope will be similar to Figure 8-9.

Multiple periodic noise sources may or may not be synchronized. If not, their fields add incoherently, and the power in their sum is equal to the sum of their individual powers. If the periodic noise sources are synchronized, however, their fields add coherently. The total power depends on the relative phase, but it may be as great as the power in a signal whose amplitude equals the sum of individual signal amplitudes.

Since the total field strength due to n coherent sources is proportional to n instead of $\sqrt{n}$, the field can become quite strong. Consequently, synchronized periodic signals should be minimized. Where they are unavoidable, as in clock distribution circuitry, we must use particular care to minimize the area a, possibly by using striplines.

Since differential-mode radiation is directly proportional to I and a and inversely proportional to $\sqrt{T_c}$ and t_r (for $f > 1/[\pi t_r]$), this radiation may be minimized as follows:

- The lowest possible power should be used, to minimize I.
- The rise times should be made as long as possible by use of devices that are no faster than necessary. Devices that exhibit sinusoidal leading and trailing edges are preferable (see Chapter 7, Problem 10).
- The lowest possible clock frequency should be used.
- The areas of current loops should be minimized. A multilayer PWC with a solid ground plane is best, but otherwise, a ground grid should be used.
- Leads carrying synchronous periodic waveforms, such as clock leads, should be minimized.
- Dedicated return paths should be used for unavoidable clock leads and other leads carrying synchronous periodic waveforms. A return conductor on each side of the PWC is preferable.
- Current should be forced to return via the path closest to the incident current, by judicious choice of ground connection points.
- Cables that need not carry high frequencies should be provided with decoupling capacitors in order to prevent high frequencies from accidentally reaching the cables.
- Shielded cables, or *balanced* twisted pairs, should be used for interconnecting cables that must carry high signal frequencies, to minimize a.

8-6 ELECTRIC FIELD SOURCE CHARACTERISTICS

Potential differences exist between conductors on printed wiring cards. As the potential differences vary, the conductors behave as numerous oscillating electric dipoles in parallel. They can therefore be sources of electric fields, as shown in Section 8-1. The length l of each dipole is the perpendicular distance between conductors. If V is the potential difference between two parallel conductors and C is the capacitance between them, then the electric field intensity due to the potential difference is

$$\mathbf{E}_s = -\frac{\beta^2 l Q e^{-j\beta r}}{4\pi\epsilon r}\, \mathbf{a}_E = -\frac{\beta^2 l C V e^{-j\beta r}}{4\pi\epsilon r}\, \mathbf{a}_E \qquad (8\text{-}145)$$

If the two conductors are parallel wires, we may use the capacitance formula from Chapter 3. If they are a stripline, or if one conductor is the ground plane on a PWC, a simpler formula results. The capacitance of a stripline is

$$C = \frac{\epsilon_r \epsilon_0 a}{l} \qquad (8\text{-}146)$$

where a is the area of each stripline conductor, l is the perpendicular distance between them, and ϵ_r is the relative permittivity of the PWC dielectric. The electric field intensity in free space outside the PWC dielectric material is then

$$\mathbf{E}_s = -\frac{\beta^2 l (\epsilon_r \epsilon_0 a/l) V e^{-j\beta r}}{4\pi\epsilon_0 r}\, \mathbf{a}_E = -\frac{\beta^2 \epsilon_r a V e^{-j\beta r}}{4\pi r}\, \mathbf{a}_E \qquad (8\text{-}147)$$

This neglects the dielectric boundary at the surface of the PWC, since we assume $\epsilon = \epsilon_0$ for the solutions to Maxwell's equations. The magnitude of the phasor is

$$|\mathbf{E}_s| = \frac{\beta^2 \epsilon_r a V}{4\pi r}\, \mathbf{a}_E = \frac{\pi f^2 \epsilon_r a V}{c^2 r}\, \mathbf{a}_E$$

$$= 3.49 \cdot 10^{-3} \cdot \frac{f^2 \epsilon_r a V}{r}\, \mathbf{a}_E \ \ \mu\text{V/m} \qquad (8\text{-}148)$$

with f in megahertz, a in square centimeters, and V in volts.

If the conductors carrying the voltage V are near a conductive plane, the reflection from its surface will add to the incident wave. As shown in the previous section, the maximum field intensity will then be double the value calculated in Equation (8-148), or

$$|\mathbf{E}_s| = \frac{2\pi f^2 \epsilon_r a V}{c^2 r}\, \mathbf{a}_E = 6.98 \cdot 10^{-3} \cdot \frac{f^2 \epsilon_r a V}{r}\, \mathbf{a}_E \ \ \mu\text{V/m} \qquad (8\text{-}149)$$

with f in megahertz, a in square centimeters, and V in volts, as before. Since EMI measurements are normally made over a conductive floor, Equation (8-149) must be used to estimate such emissions.

To estimate a typical field strength, we may substitute values into Equation (8-149). For example, for a 3-V, 50-MHz signal on a stripline 10 cm long and 1 mm wide, on a PWC with $\epsilon_r = 4$, the field intensity 10 m away is

$$|\mathbf{E}_s| = 6.98 \cdot 10^{-3} \left(\frac{50^2 \cdot 4 \cdot 10 \cdot 0.1 \cdot 3}{10}\right) \mathbf{a}_E = 20.9\, \mathbf{a}_E \ \ \mu\text{V/m} \qquad (8\text{-}150)$$

The electric field sources discussed above depend only on potential difference and can exist even with no current flow. For voltage levels used in solid-state circuits, potential differences are usually less troublesome than current loops. However, we should not ignore them. Their frequency response is comparable to that for current loops, since Equations (8-149) and (8-130) contain the same function of frequency. Due to the analogy between electric and magnetic fields between conductors, the preventive measures recommended earlier for current-loop sources also work well for the potential-difference sources discussed above.

A more serious source of electric fields is common-mode currents. To produce an oscillating dipole, an alternating current $i(t)$ must flow through a conductor between the two ends of the dipole. Then the charge $q(t)$ builds up and discharges at each end of the dipole according to the following relation:

$$i(t) = \frac{dq(t)}{dt} \tag{8-151}$$

Using phasor notation, this becomes

$$I = j\omega Q \tag{8-152}$$

If the current flows through a perfect conductor in a closed loop that is much smaller than $\lambda/(2\pi)$, the voltage and current are constant everywhere along its path. The dipoles then cancel each other, and there is no net charge or voltage difference on the loop. Thus, no net electric field results from the dipoles. So a small current loop on a printed wiring card can act only as a magnetic field source, not an electric one. However, current can flow in a wire that leaves the PWC and can then return as displacement current (time-varying electric flux) through space. This flux represents the electric field generated by the dipoles. This flux is the only possible electric field source due to a flowing current. It is possible only with leads that are long enough to allow the current to vary along their length, which implies a length greater than $\lambda/(2\pi)$. The usual source is common-mode currents flowing on cables that interconnect two or more cabinets. We shall now investigate the fields due to such common-mode currents.

An oscillating current I in a wire of length l (where $l \ll \lambda/(2\pi)$ so that I is constant over the length of wire), is equivalent to an oscillating electric dipole lQ. The far field due to this current is equivalent to that caused by the dipole, and is

$$\mathbf{E}_s = -\frac{\omega^2 \mu l Q e^{-j\beta r}}{4\pi r} \mathbf{a}_E$$

$$= \frac{j\omega \mu I l e^{-j\beta r}}{4\pi r} \mathbf{a}_E \tag{8-153}$$

Its time magnitude $|\mathbf{E}_s|$ is then

$$|\mathbf{E}_s| = \frac{\omega \mu |I| l}{4\pi r} \mathbf{a}_E = \frac{f\mu |I| l}{2r} \mathbf{a}_E \tag{8-154}$$

If the conductor carrying the current I is near a conductive plane, the reflection will add to the incident wave. The maximum field intensity will then be double the value calculated in Equation (8-154), or

$$|\mathbf{E}_s| = \frac{f\mu|I|l}{r}\, \mathbf{a}_E \qquad (8\text{-}155)$$

Since EMI measurements are normally made over a conductive floor, Equation (8-155) must be used to estimate such emissions. In free space, the field intensity is

$$|\mathbf{E}_s| = \frac{f\mu_0|I|l}{r}\, \mathbf{a}_E = \frac{1.26\ \mu\text{H/m} \cdot f|I|l}{r}\, \mathbf{a}_E \qquad (8\text{-}156)$$

Equation (8-156) is normally used to estimate the field strength at a known distance from an electric-field source.

A workable example is instructive but not very realistic. Consider a 10-MHz sinusoidal current of 1-mA peak, flowing in a 20-cm length of wire connected between two spheres, each 2 cm in diameter, as shown in Figure 8-10. The current may be generated, for example, by a small battery-powered oscillator somewhere on the wire. At 10 MHz, $\lambda/(2\pi) \approx 5$ m $\gg$ 20 cm, so the current is effectively constant along the wire. This dipole is of practical size and is not infinitesimal, but it is small compared to $\lambda/(2\pi)$, so our earlier analysis is still valid. The average current is 0.636 mA, and in one quarter cycle, or 25 ns, a charge of ±0.636 mA $\cdot$ 25 ns, or ±15.9 pC must flow onto each sphere. The capacitance of each sphere is $4\pi\epsilon_0 r$, or 1.11 pF, so the peak voltage between the spheres is 14.3 V, a reasonable value. This is an electric-field source due to the 1-mA current, and the peak field intensity 10 m away is

$$|\mathbf{E}_s| = \frac{1.26\ \mu\text{H/m} \cdot 10\ \text{MHz} \cdot 1\ \text{mA} \cdot 20\ \text{cm}}{10\ \text{m}}\, \mathbf{a}_E = 251.3\ \mathbf{a}_E\ \mu\text{V/m} \qquad (8\text{-}157)$$

Figure 8-10   Current flowing between two spheres with time-varying charges

Unless the charge builds up at the ends of the dipole, the dipole cannot act as an electric-field source. If we place several such dipoles end-to-end, as shown in Figure 8-11, then if all currents are equal and in phase, all charges cancel out except at the two ends. The dipoles become equivalent to one long dipole, and the field intensity is the sum of the fields due to the individual dipoles. If the currents are not equal, then we must treat each dipole separately, but the total field intensity is still the sum of the individual fields. Smaller charges then continuously build up and discharge at the ends of each dipole. In the limit, the dipoles become infinitesimal, and the charge distribution becomes continuous. If we know the current I as a function of position along the wire, we can integrate Equation (8-156) over the length of the wire to obtain the total field intensity. We will use this procedure when we study antennas in Chapters 10 and 11.

In a practical application, there are usually no spheres or other large surfaces on the wire, but the charge becomes distributed on the wire itself. For this to occur, the current I must vary along its path. Usually, I is common-mode current flowing in cables, which should ideally be zero. Unlike the voltage and current between components on a PWC, the common-mode current flowing into a cable and returning through space is difficult to calculate. This requires application of antenna theory to the cable geometry, which we will discuss in Chapter 11. We can, however, measure the current with a high-frequency current probe. We observe that field intensity is proportional to the frequency of the source of the current I. With this knowledge, we can predict the frequency characteristics of the radiated field.

To obtain the spectrum of the field due to a random trapezoidal common-mode current, we substitute the current I from Equation (8-133) into Equation (8-156). The resulting field intensity is

$$|\mathbf{E}_s(f, \Delta f)| = \frac{f\mu_0 I_0 l}{r} \sqrt{\frac{T_c \Delta f}{2}} \left| \frac{\sin(\pi f T_c)}{\pi f T_c} \frac{\sin(\pi f t_r)}{\pi f t_r} \right| \mathbf{a}_E$$

$$= \frac{\mu_0 I_0 l}{\pi r} \sqrt{\frac{\Delta f}{2 T_c}} \left| \frac{\sin(\pi f T_c)\sin(\pi f t_r)}{\pi f t_r} \right| \mathbf{a}_E$$

$$= \left(\frac{\mu_0}{\pi\sqrt{2}}\right) \frac{I_0 l}{r} \sqrt{\frac{\Delta f}{T_c}} \left| \frac{\sin(\pi f T_c)\sin(\pi f t_r)}{\pi f t_r} \right| \mathbf{a}_E \qquad \textbf{(8-158)}$$

Let K_{com} represent the parenthesized constant in Equation (8-158), or

$$K_{com} = \frac{\mu_0}{\pi\sqrt{2}} = 2.83 \cdot 10^{-7} \frac{\text{H}}{\text{m}} = 2.83 \cdot 10^{-7} \cdot \frac{\text{V s}}{\text{A m}} = 2.83 \cdot 10^5 \frac{\mu\text{V}\,\mu\text{s}}{\text{A m}} \qquad \textbf{(8-159)}$$

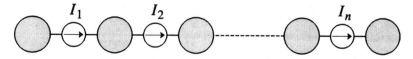

Figure 8-11 Varying current represented by multiple dipoles

Then, with I_0 expressed in amperes, r and l in meters, f and Δf in megahertz, and T_c and t_r in microseconds, Equation (8-158) becomes

$$|\mathbf{E}_s(f, \Delta f)| = K_{com}\frac{I_0 l}{r}\sqrt{\frac{\Delta f}{T_c}}\left|\frac{\sin(\pi f T_c)\sin(\pi f t_r)}{\pi f t_r}\right|\mathbf{a}_E$$

$$= 2.83\cdot 10^5 \cdot \frac{I_0 l}{r}\sqrt{\frac{\Delta f}{T_c}}\left|\frac{\sin(\pi f T_c)\sin(\pi f t_r)}{\pi f t_r}\right|\mathbf{a}_E\ \mu V/m \qquad \textbf{(8-160)}$$

We now use reasoning similar to that used for the current loop. For common-mode sources, for $f \leq 1/(\pi T_c)$, the upper bound of $|\mathbf{E}_s|$ is proportional to f and thus increases 20 dB per decade. For frequencies in the interval $1/(\pi T_c) \leq f \leq 1/(\pi t_r)$, since $\sin(\pi f T_c) \leq 1$ and $\sin(\pi f t_r) \leq \pi f t_r$, it follows that

$$|\mathbf{E}_s(f, \Delta f)| \leq K_{com}\frac{I_0 l}{r}\sqrt{\frac{\Delta f}{T_c}}\,\mathbf{a}_E \qquad \textbf{(8-161)}$$

In this interval, $|\mathbf{E}|$ varies sinusoidally with frequency, but its upper bound is constant. For $f \geq 1/(\pi t_r)$, $|\mathbf{E}_s|$ still varies sinusoidally with frequency, and its upper bound is proportional to $1/f$ and thus *decreases* 20 dB per decade. From Equations (8-159) and (8-161), if $I_0 = 1$ A, $l = 1$ cm, $r = 1$ m, $T_c = 1$ μs, and $t_r = 0.1$ μs, the upper bound of $|\mathbf{E}_s|$ over the mid-range frequencies is

$$\frac{|\mathbf{E}_s(f, \Delta f)|}{\sqrt{\Delta f}} \leq \frac{K_{com}I_0 l}{r\sqrt{T_c}}\,\mathbf{a}_E = \frac{2.83\cdot 10^5 \cdot 1 \cdot 0.01}{1\cdot\sqrt{1}}\frac{\mu V}{m\sqrt{MHz}}\cdot\mathbf{a}_E$$

$$\leq 2.83\cdot 10^3 \cdot \frac{\mu V}{m\sqrt{MHz}}\,\mathbf{a}_E \qquad \textbf{(8-162)}$$

To express this field strength in dB$_{\mu V/m/MHz}$, we again use the definition from Equation (7-23):

$$E_{dB\mu V/m/MHz} = 10\log\frac{|\mathbf{E}_s(f, \Delta f)|^2/\Delta f}{1\ \mu V^2/m^2/MHz}$$

$$\leq 10\log[(2.83\cdot 10^3)^2] = 69.03\ dB_{\mu V/m/MHz} \qquad \textbf{(8-163)}$$

Its spectral density plot, for I_0, l, r, T_c, and t_r as given above, appears in Figure 8-12, where $f_c = 1/T_c$ is the clock frequency. Comparing this plot to Figure 8-9, we note that radiation due to common-mode currents is dominant at lower frequencies. The radiation efficiency of common-mode current paths is greater at those frequencies.

Common-mode radiation is directly proportional to I and l while being inversely proportional to $\sqrt{T_c}$ and t_r (for $f > 1/[\pi t_r]$). Thus, most of the techniques

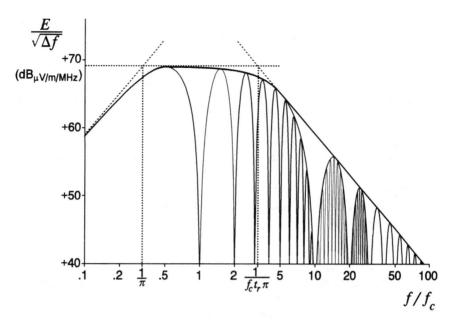

Figure 8-12   Spectral density of radiation 1 m away from a 1-ampere common-mode current source of 1-cm length, with $T_c = 1$ μs and $t_r = 0.1$ μs.

mentioned in Section 8.5 that reduce differential-mode radiation are also effective for common-mode. In addition, the following steps are advisable:

- Cable lengths should be minimized.
- The source voltage driving each cable should be minimized.
- Common-mode current should be bypassed through a capacitor to ground, at a point as near as possible to the current source. This is possible only if the cable does not intentionally carry high-frequency differential-mode currents. The ground must be free of noise and must have a low impedance. Bypassing will be discussed further in Chapter 11.
- A common-mode choke should be inserted in each cable, or preferably, a balanced circuit should be used. Common-mode chokes are effective if they can provide a high enough impedance, usually 1 kΩ or higher. It is usually necessary to bulkhead-mount the choke to prevent displacement current from flowing through the free space around it.
- The cable should be shielded, and the other end must then be protected also; otherwise, the problem merely moves to the other end of the cable.

8-7 PROBLEMS

1. Show that Equations (8-15) and (8-16) satisfy all components of Maxwell's equations.
2. Compute the electric field spectral density, $E/\sqrt{\Delta f}$, of a random digital current $I_0 = 1$ ampere, flowing in a loop whose area is 1 cm^2, at the frequency $f = 0.1f_c$, at a distance

$r = 1$ m. A field strength of 0 dB$_{\mu V/m/MHz}$ corresponds to 1 $\mu V/m/MHz^{1/2}$. Compute the number of decibels by which $E/\sqrt{\Delta f}$ is below this reference level, and thus show that Figure 8-9 is scaled correctly.

3. A printed wiring card contains 200 current loops, each carrying a random binary signal unrelated to the others. In each loop, a logic "0" corresponds to a current of 0.15 mA, and a logic "1" corresponds to a current of zero. The average conductor length is 10 cm, and on average, the return current paths are 1 cm away from the incident current. The clock frequency is 10 MHz, the pulses are trapezoidal, and their rise and fall times are 10 ns. The card is located 1 m above a conductive floor. Calculate the estimated electric field spectral density at 5 MHz due to all 200 current loops, at a distance of 3 m from the card.

4. Express the answer to Problem 3 in dB$_{\mu V/m/MHz}$.

5. The clock circuitry on the printed wiring card in Problem 3 carries a 10-MHz periodic trapezoidal wave of 50 percent duty cycle. It uses an average current of 10 mA, flowing in loops whose areas total 30 cm^2. What is the field intensity at 10 MHz due to the clock circuitry, at a distance of 3 m from the card?

6. The circuit of Problem 3 is now built on a card containing a ground plane, as discussed in Chapter 6. All parameters remain the same except that the return paths are now 0.4 mm from the incident paths. Calculate the average electric field intensity.

7. A cable 1 m long carries a random digital signal generated by a 10-MHz clock. The peak common-mode current is measured and found to average 10 mA along the cable. The rise and fall times are 10 ns. Calculate the electric field intensity at a distance of 3 m from the cable.

8. Derive the formula for the electric field intensity due to a periodic common-mode current consisting of trapezoidal pulses.

9

CABINET SHIELDING
TO REDUCE
RADIATED COUPLING

It is always best to reduce noise emissions at the source, using the methods of Chapter 8. When the source radiation or receptor sensitivity cannot be sufficiently reduced using those techniques, however, shielding is necessary. We shall see that it is easy to shield against most types of electromagnetic radiation, unless a hole in the shield is necessary for some reason. Holes in a shield cause leakage, which will be discussed in Chapters 10 and 11. In the present chapter we shall investigate the processes by which a solid shield attenuates radiated electromagnetic fields.

A time-varying electromagnetic field irradiating a conductive shield causes a current to flow in the shield, which generates a new field. The direction of the new field is such that it subtracts from original field on the side of the shield opposite the approaching wave. Thus, the shield attenuates the original field as it passes through. The amount of attenuation depends on the current, which in turn depends on the field strength and the shield characteristics. We will now study these effects quantitatively, using techniques originally developed by Schelkunoff.[1]

A shield affects an electromagnetic field in three ways. Some electromagnetic energy is absorbed into the shield material and is dissipated as heat due to its resistance. Other energy is reflected from the surface of the shield due to the impedance mismatch and never penetrates the shield. Finally, the field may be redirected due to the magnetic permeability of the shield. Each of these phenomena will be studied separately.

---

1. S. A. Schelkunoff, *Electromagnetic Waves* (New York, N.Y.: D. Van Nostrand, Inc., 1943), pp. 303–6.

9-1 ABSORPTION LOSSES IN NONMAGNETIC SHIELDS

Maxwell's equations govern electric and magnetic fields not only in free space, but also inside conductors, such as shield material. For a plane wave in a material of conductivity σ, permeability μ, and permittivity ϵ, their solutions in phasor form are

$$\mathbf{E}_s = -\omega^2 \mu K \exp\left[-j\omega r \sqrt{\mu\left(\epsilon + \frac{\sigma}{j\omega}\right)}\right] \mathbf{a}_E \tag{9-1}$$

$$\mathbf{H}_s = -\omega^2 \sqrt{\mu\left(\epsilon + \frac{\sigma}{j\omega}\right)} K \exp\left[-j\omega r \sqrt{\mu\left(\epsilon + \frac{\sigma}{j\omega}\right)}\right] \mathbf{a}_H \tag{9-2}$$

for some constant K, which depends on the source. These solutions are the same as for a nonconductive material as demonstrated in Chapter 8, except that $\epsilon + \sigma/j\omega$ replaces ϵ and $1/\sqrt{\mu(\epsilon + \sigma/j\omega)}$ replaces c. In place of $j\beta$, we define the *propagation constant* $\gamma = \sqrt{j\omega\mu(\sigma + j\omega\epsilon)}$. With this definition, Equations (9-1) and (9-2) become

$$\mathbf{E}_s = -\omega^2 \mu K e^{-\gamma r} \mathbf{a}_E \tag{9-3}$$

$$\mathbf{H}_s = j\omega\gamma K e^{-\gamma r} \mathbf{a}_H \tag{9-4}$$

The imaginary part of γ is the change of phase of $\mathbf{E}_s$ or $\mathbf{H}_s$ per unit of distance r. It is therefore the phase constant β defined in Chapter 8, but we will find that its value is much larger in a conductor than in free space. Let α be the real part of γ so that $\gamma = \alpha + j\beta$; then Equations (9-3) and (9-4) become

$$\mathbf{E}_s = E_0 e^{-\alpha r} e^{-j\beta r} \mathbf{a}_E \tag{9-5}$$

$$\mathbf{H}_s = E_0 \frac{\gamma}{j\omega\mu} e^{-\alpha r} e^{-j\beta r} \mathbf{a}_H \tag{9-6}$$

where

$$E_0 = -\omega^2 \mu K \tag{9-7}$$

Note that if $\sigma = 0$, then $\gamma = \sqrt{j\omega\mu \cdot j\omega\epsilon} = j\omega\sqrt{\mu\epsilon} = j\beta$ and Equations (9-5) and (9-6) reduce to Equations (8-77) and (8-78), as they should. The time magnitudes of the phasors $\mathbf{E}_s$ and $\mathbf{H}_s$ are

$$|\mathbf{E}_s| = E_0 e^{-\alpha r} \mathbf{a}_E \tag{9-8}$$

$$|\mathbf{H}_s| = E_0 \frac{\gamma}{j\omega\mu} e^{-\alpha r} \mathbf{a}_E \tag{9-9}$$

Therefore, as the wave penetrates the conductor, the amplitudes of both the $\mathbf{E}$ and the $\mathbf{H}$ fields decrease exponentially. The process is known as *absorption loss*. The quantity α is the natural logarithm of the change in amplitude per unit of distance r. It is called the *attenuation constant*.

    Consider a shield of thickness t. Let $\mathbf{E}_1$ be the intensity of the electric field just after entering the shield material at surface 1. Let $\mathbf{E}_{1a}$ be the intensity of the attenuated field that has passed through the shield and is about to leave the shield material at surface 2. The absorption loss A caused by this shield, expressed in decibels, is

$$A = 20 \log \frac{\|\mathbf{E}_1\|}{\|\mathbf{E}_{1a}\|} = 20 \log \frac{1}{e^{-\alpha t}} = 20 \, (\log e) \, \alpha t$$

$$= 8.686 \text{ dB} \cdot \alpha t \qquad (9\text{-}10)$$

At the distance $1/\alpha$ into the shield material, the fields have decreased to $1/e$ of their values just inside the boundary. Let δ represent this distance:

$$\delta = \frac{1}{\alpha} \qquad (9\text{-}11)$$

A current $\mathbf{J} = \sigma \mathbf{E}$ flows through the conductor, and at the distance δ, the current also decreases to $1/e$ of its value at the surface. This is called *skin effect*, and the distance δ is the *skin depth*. From Equation (9-10),

$$A = 20 \, (\log e) \, \frac{t}{\delta} = 8.686 \text{ dB} \cdot \frac{t}{\delta} \qquad (9\text{-}12)$$

Thus, the absorption loss due to a shield of thickness δ is 8.686 dB. This amount of attenuation is also known as 1 neper. Thus, any shield provides 1 neper of attenuation, or 8.686 dB, per skin depth of its thickness.

For good conductors, $\sigma \gg j\omega\epsilon$ and therefore ϵ is negligible. Then the propagation constant γ becomes

$$\gamma = \sqrt{j\omega\mu\sigma} = \sqrt{\omega\mu\sigma} \, \sqrt{j} = \sqrt{\omega\mu\sigma} \, \frac{1+j}{\sqrt{2}} = \sqrt{\frac{\omega\mu\sigma}{2}} \, (1+j) \qquad (9\text{-}13)$$

Thus $\alpha = \beta$, and both are given by

$$\alpha = \beta = \sqrt{\frac{\omega\mu\sigma}{2}} \qquad (9\text{-}14)$$

Therefore, for good conductors,

$$\delta = \frac{1}{\alpha} = \sqrt{\frac{2}{\omega\mu\sigma}} = \frac{1}{\beta} = \frac{\lambda}{2\pi} \qquad (9\text{-}15)$$

We will later note that λ is much shorter in a conductor than in free space. From Equation (9-10), we may express A simply in terms of σ, μ, and f:

$$A = 8.686 \text{ dB} \cdot \sqrt{\frac{\omega\mu\sigma}{2}} \, t$$

$$= 8.686 \text{ dB} \cdot \sqrt{\frac{2\pi}{2}} \, t\sqrt{f\mu\sigma}$$

$$= 15.39 \text{ dB} \cdot t\sqrt{f\mu\sigma} \qquad (9\text{-}16)$$

We also may express δ and A in terms of the relative permeability μ_r and relative conductivity σ_r of the shield material. Relative conductivity is defined with respect to annealed copper, for which the conductivity, designated σ_{Cu}, is $5.82 \cdot 10^7$ Ʊ/m. The relative values are $\sigma_r = \sigma_{Cu}$ and $\mu_r = \mu/\mu_0$, where $\mu_0 = 1.257$ μH/m. Note that, while $\epsilon_r \geq 1$ and $\mu_r \geq 1$ for most materials, $\sigma_r \leq 1$ for common material 3 except silver. In terms of σ_r, μ_r, and f,

$$\delta = \sqrt{\frac{2}{(2\pi f)(\mu_r \mu_0)(\sigma_r \sigma_{Cu})}} = \frac{\sqrt{2/(2\pi\mu_0\sigma_{Cu})}}{\sqrt{f\mu_r\sigma_r}} = \frac{0.066}{\sqrt{f\mu_r\sigma_r}} \text{ mm} \qquad (9-17)$$

and

$$A = 15.395 \text{ dB} \cdot t\sqrt{\mu_0\sigma_{Cu}} \sqrt{f\mu_r\sigma_r}$$

$$= 132 \text{ dB} \cdot t\sqrt{f\mu_r\sigma_r} \qquad (9-18)$$

with the frequency f in megahertz and the thickness t in millimeters.

For example, consider a shield made of tempered aluminum, for which $\sigma_r = 0.4$. Since tempered aluminum is nonmagnetic, $\mu_r = 1$. Therefore, for a frequency of 1 MHz, the skin depth is

$$\delta = \frac{0.066}{\sqrt{1 \cdot 1 \cdot 0.4}} \text{ mm} = 0.104 \text{ mm} \qquad (9-19)$$

The wavelength inside the aluminum conductor is 2π times this value, or 0.655 mm. In good conductors, the skin depth and wavelength are very small compared to the wavelength in free space. A tempered aluminum shield of thickness 1 mm, or approximately 10 δ, with no holes, would attenuate a 1-MHz plane wave by approximately 83 dB. This considers only absorption loss.

9-2 REFLECTION LOSSES DUE TO NONMAGNETIC SHIELDS

A wave approaching a shield is also *reflected* due to the impedance mismatch at the shield surface. Unlike absorption loss, reflection does not annihilate the undesired field but only redirects it, hopefully in a direction where it will do less harm. This effect is in addition to the absorption loss discussed in the preceding section.

The simplest case of a reflection is a plane wave approaching a flat shield perpendicularly. Then both $\mathbf{E}$ and $\mathbf{H}$ are parallel to the shield surface. Let Z_w be the wave impedance of the plane wave in free space. From Chapter 8 this is

$$Z_w = \frac{|\mathbf{E}_0|_s}{|\mathbf{H}_0|_s} = \sqrt{\frac{\mu_0}{\epsilon_0}} = \eta_0 \qquad (9-20)$$

where $\mathbf{E}_0$ and $\mathbf{H}_0$ are the fields that are outside the shield, and $|\mathbf{E}_0|_s$ and $|\mathbf{H}_0|_s$ are phasors representing their magnitudes. The ratio of $|\mathbf{E}|_s$ to $|\mathbf{H}|_s$ within the metal of the shield is called the *shield impedance*, designated Z_s. From Equations (9-1) and (9-2), the shield impedance for a good conductor is equal to

$$Z_s = \frac{|\mathbf{E}_1|_s}{|\mathbf{H}_1|_s} = \frac{\mu}{\sqrt{\mu\left(\epsilon + \dfrac{\sigma}{j\omega}\right)}} = \sqrt{\frac{j\omega\mu}{\sigma + j\omega\epsilon}}$$

$$\approx \sqrt{\frac{j\omega\mu}{\sigma}} = \sqrt{\frac{\omega\mu}{\sigma}} \angle \frac{\pi}{4} \tag{9-21}$$

where $\mathbf{E}_1$ and $\mathbf{H}_1$ are the fields within the shield. The shield impedance of tempered aluminum at 1 MHz, for example, is

$$Z_s = \sqrt{\frac{1\ \text{MHz} \cdot 0.4\pi\ \mu\text{H/m}}{2.33 \cdot 10^7 \mho/\text{m}}} \angle 45° = 2.32 \cdot 10^{-4}\ \Omega \angle 45° \tag{9-22}$$

For any good conductor, Z_s is a small fraction of an ohm, and its angle is always 45°.

To study the $\mathbf{H}$ field at the shield surface, we use the closed path of integration shown in Figure 9-1. The path is a narrow rectangle whose longest sides are parallel to $\mathbf{H}_0$, which is parallel to the shield surface. Half of the path is buried in the solid metal of the shield. Let the width of the rectangle approach zero; then the enclosed area also approaches zero. As mentioned earlier, current is flowing in the shield perpendicular to $\mathbf{H}_0$. Although the current layer is very thin due to skin effect, it still has finite thickness and is not a true "sheet" of current. Its density $\mathbf{J}$ is everywhere finite. Therefore, as the area outlined by the path in Figure 9-1 approaches zero, so does the current enclosed by the path. Thus, by Ampere's law, the line integral of $\mathbf{H}$

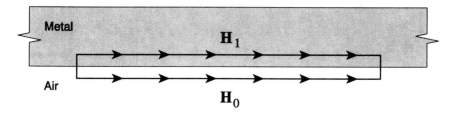

Figure 9-1   Closed path of integration through $\mathbf{H}_0$ and $\mathbf{H}_1$

around the path must approach zero. The positions of the ends of the rectangle are arbitrary, and the integral must approach zero regardless of their positions. This is possible only if H_1 at every point is equal to H_0 at the directly opposite point on the path, since only then will the two always cancel. This implies that for H_0 parallel to the shield surface, $H_1 = H_0$ everywhere at the surface.

The E field is also parallel to the shield surface, and similar reasoning implies that $E_1 = E_0$. If $Z_s \neq Z_w$, these four fields alone cannot simultaneously satisfy these two boundary conditions. To satisfy them for $Z_w > Z_s$ requires additional E and H fields in directions that increase the total H field while reducing the total E field. Let E_r and H_r be these additional fields, as shown in Figure 9-2. Now, due to the respective directions of E_r and H_r, the boundary conditions become

$$(E_1)_s = (E_0)_s + (E_r)_s = (|E_0|_s - |E_r|_s)a_E \qquad (9\text{-}23)$$

$$(H_1)_s = (H_0)_s + (H_r)_s = (|H_0|_s + |H_r|_s)a_H \qquad (9\text{-}24)$$

where a_E and a_H are unit vectors in the directions of E_0 and H_0, respectively.

The new fields also must satisfy Maxwell's equations. Since E_r and H_r are in the free space outside the shield material, this requires that

$$Z_w = \frac{|E_r|_s}{|H_r|_s} \qquad (9\text{-}25)$$

We also note that due to the directions of E_r and H_r, the vector $P_r = E_r \times H_r$ points in the direction opposite to P_0 and P_1. Therefore, the new fields form a wave

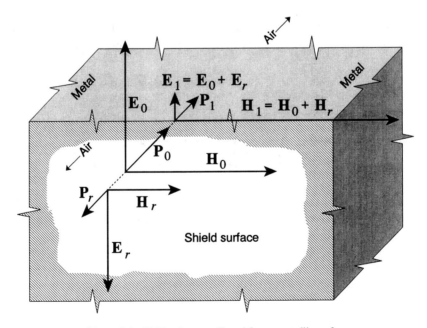

Figure 9-2   Fields of wave reflected from a metallic surface

traveling in the $-r$ direction, opposite to the other waves. This wave is a reflection from the boundary surface. Equations (9-23), (9-24), and (9-25) are necessary and sufficient conditions to determine E_1 and H_1 uniquely. Fields E_1 and H_1 are called the *transmitted fields*.

We obtain the transmitted field intensities from the definitions of Z_w and Z_s as follows. For simplicity, let $E_0 = |E_0|_s$, and similarly for the other fields. Then

$$Z_w = \frac{E_r}{H_r} = \frac{E_0 - E_1}{H_1 - H_0} = \frac{E_0 - E_1}{E_1/Z_s - E_0/Z_w}$$

$$E_1 Z_w - E_0 Z_s = E_0 Z_s - E_1 Z_s$$

$$E_1 = E_0 \frac{2Z_s}{Z_w + Z_s} \tag{9-26}$$

$$H_1 = \frac{E_1}{Z_s} = E_0 \frac{2}{Z_w + Z_s} = H_0 \frac{2Z_w}{Z_w + Z_s} \tag{9-27}$$

The *transmission coefficient*, designated T, is the ratio of the transmitted field E_1 to the original incident field E_0. Its value is

$$T = \frac{E_1}{E_0} = \frac{2Z_s}{Z_w + Z_s} = \frac{2\sqrt{j\omega\mu/\sigma}}{\sqrt{\mu_0/\epsilon_0} + \sqrt{j\omega\mu/\sigma}}$$

$$= \frac{2\sqrt{j\omega\mu_r\epsilon_0/\sigma}}{1 + \sqrt{j\omega\mu_r\epsilon_0/\sigma}} \tag{9-28}$$

The ratio of the corresponding H fields is related to T and is

$$\frac{H_1}{H_0} = \frac{2Z_w}{Z_w + Z_s} = \frac{2\sqrt{\mu_0/\epsilon_0}}{\sqrt{\mu_0/\epsilon_0} + \sqrt{j\omega\mu/\sigma}}$$

$$= \frac{2}{1 + \sqrt{j\omega\mu_r\epsilon_0/\sigma}} = 2 - T \tag{9-29}$$

Note that since $Z_w > Z_s$, although $E_1 < E_0$, $H_1 > H_0$. Therefore, the magnetic field *gains* strength as the wave enters the shield material through surface 1. It will weaken again when the wave leaves the shield material through surface 2.

Now we can combine the effects of absorption and reflection. A cross-sectional view of a shield appears in Figure 9-3, which also shows the P vectors indicating the direction of travel of the waves. For clarity, the figure shows the waves approaching the shield at a slightly oblique angle, but they actually meet surface 1 of the shield perpendicularly. At surface 2, but still within the shield, the field strengths have been reduced by absorption loss and their new values, designated E_{1a} and H_{1a}, are

$$E_{1a} = E_1 e^{-t/\delta} e^{-jt/\delta} = E_0 T e^{-t/\delta} e^{-jt/\delta}$$

$$= E_0 \frac{2Z_s}{Z_w + Z_s} e^{-t/\delta} e^{-jt/\delta} \tag{9-30}$$

The boundary conditions at surface 2 cause another reflection, as shown in Figure 9-4. The transmitted fields, which escape surface 2 and pass completely through the shield, are designated E_{1t} and H_{1t}. The reflected fields are designated E_{1r} and H_{1r}. The wave and shield impedances imply that the reflected fields are

$$Z_w = \frac{E_{1t}}{H_{1t}} = \frac{E_{1a} + E_{1r}}{H_{1a} - H_{1r}} = \frac{E_{1a} + E_{1r}}{(E_{1a} - E_{1r})/Z_s}$$

$$Z_w E_{1a} - Z_w E_{1r} = Z_s E_{1a} - Z_s E_{1r}$$

$$E_{1r} = E_{1a} \frac{Z_w - Z_s}{Z_w + Z_s} \tag{9-31}$$

$$H_{1r} = \frac{E_{1r}}{Z_s} = \frac{E_{1a}}{Z_s} \cdot \frac{Z_w - Z_s}{Z_w + Z_s} = H_{1a} \frac{Z_w - Z_s}{Z_w + Z_s} \tag{9-32}$$

The transmitted fields are given by

$$E_{1t} = E_{1a} + E_{1r} = E_{1a}\left(1 + \frac{Z_w - Z_s}{Z_w + Z_s}\right) = E_{1a} \frac{2Z_w}{Z_w + Z_s} \tag{9-33}$$

$$H_{1t} = H_{1a} - H_{1r} = H_{1a}\left(1 - \frac{Z_w - Z_s}{Z_w + Z_s}\right) = H_{1a} \frac{2Z_s}{Z_w + Z_s} \tag{9-34}$$

The external reflection, $E_r \times H_r$, leaves the shield and never can penetrate it. The internal reflection $E_{1r} \times H_{1r}$, however, is again reflected off the inside of surface 1 and again reaches surface 2. The process repeats, and as each reflection reaches either surface, some field energy escapes. All energy that escapes through

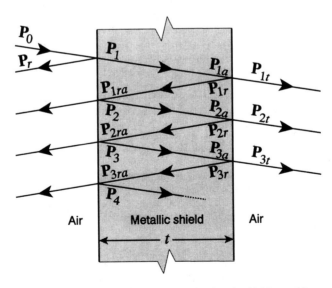

Figure 9-3 Wave propagation and reflections in shield material

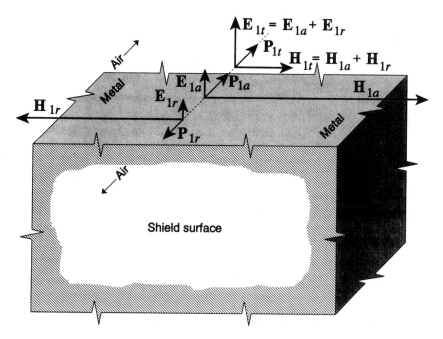

Figure 9-4   Fields of wave reflected inside a metallic shield

surface 2, opposite the original source, effectively passes through the shield. Let Γ be the fraction of the field that is internally reflected at either surface, defined as the *reflection coefficient*:

$$\Gamma = \frac{E_{1r}}{E_{1a}} = \frac{H_{1r}}{H_{1a}} = \frac{Z_w - Z_s}{Z_w + Z_s} \tag{9-35}$$

After $\mathbf{E}_{1r}$ and $\mathbf{H}_{1r}$ have propagated through the shield, been reflected at surface 1, and again approached surface 2, their field strengths are

$$E_{2a} = E_2 e^{-t/\delta} e^{-jt/\delta}$$

$$= E_{1ra} \Gamma e^{-t/\delta} e^{-jt/\delta}$$

$$= E_{1a} (\Gamma e^{-t/\delta} e^{-jt/\delta})^2 \tag{9-36}$$

$$H_{2a} = H_{1a} (\Gamma e^{-t/\delta} e^{-jt/\delta})^2 \tag{9-37}$$

This wave causes another transmitted wave and another reflected wave:

$$E_{2t} = E_{2a} \frac{2Z_w}{Z_w + Z_s} \tag{9-38}$$

$$H_{2t} = H_{2a} \frac{2Z_s}{Z_w + Z_s} \tag{9-39}$$

The process is repeated indefinitely, and the *total* transmitted electric fields are

$$E_t = E_{1t} + E_{2t} + E_{3t} + \ldots$$

$$= (E_{1a} + E_{2a} + E_{3a} + \ldots)\frac{2Z_w}{Z_w + Z_s}$$

$$= E_{1a}[1 + (\Gamma e^{-t/\delta}e^{-jt/\delta})^2 + (\Gamma e^{-t/\delta}e^{-jt/\delta})^4 + \ldots]\frac{2Z_w}{Z_w + Z_s}$$

$$= E_0\frac{2Z_s}{Z_w + Z_s}e^{-t/\delta}e^{-jt/\delta} \cdot [1 + (\Gamma e^{-t/\delta}e^{-jt/\delta})^2 + (\Gamma e^{-t/\delta}e^{-jt/\delta})^4 + \ldots]\frac{2Z_w}{Z_w + Z_s}$$

$$= E_0e^{-t/\delta}e^{-jt/\delta} \cdot \frac{4Z_wZ_s}{(Z_w + Z_s)^2} \cdot \frac{1}{1 - (\Gamma e^{-t/\delta}e^{-jt/\delta})^2} \tag{9-40}$$

Since $|e^{-jt/\delta}| = 1$, the attenuation due to the shield is

$$\frac{|E_t|}{|E_0|} = e^{-t/\delta} \cdot \frac{4|Z_w||Z_s|}{|Z_w + Z_s|^2} \cdot \frac{1}{|1 - \Gamma^2 e^{-2t/\delta}e^{-j2t/\delta}|} \tag{9-41}$$

For good conductors, $|Z_s| \ll |Z_w|$. Then

$$\Gamma = \frac{Z_w - Z_s}{Z_w + Z_s} \approx \frac{Z_w}{Z_w} = 1 \tag{9-42}$$

and

$$\frac{4|Z_w||Z_s|}{|Z_w + Z_s|^2} \approx \frac{4|Z_w||Z_s|}{|Z_w|^2} = \frac{4|Z_s|}{|Z_w|} \tag{9-43}$$

Then

$$\frac{|E_t|}{|E_0|} = e^{-t/\delta} \cdot \frac{4|Z_s|}{|Z_w|} \cdot \frac{1}{|1 - e^{-2t/\delta}e^{-j2t/\delta}|} \tag{9-44}$$

Equation (9-44) shows three loss factors: the absorption loss, the reflection loss, and a factor to account for multiple reflections. We may calculate them separately and multiply them to obtain the total loss. However, it is usually more convenient to express them in decibels and add them. We have already expressed the absorption loss in decibels as A in Equations (9-12), (9-16), and (9-18). Let R and M be the other two factors in Equation (9-44), expressed in decibels. Their formulas are

$$R = -20 \log \frac{4|Z_s|}{|Z_w|} = 20 \log \frac{|Z_w|}{4|Z_s|} \text{ dB} \tag{9-45}$$

$$M = 20 \log |1 - e^{-2\delta/t}e^{-j2\delta/t}| \text{ dB} \tag{9-46}$$

We may write Equation (9-44) in terms of A, R, and M as

$$\frac{|E_t|}{|E_0|} = 10^{-A/20} \cdot 10^{-R/20} \cdot 10^{-M/20} \tag{9-47}$$

From Equation (9-45), we may express R directly in terms of shield conductivity, permeability, and frequency. Since $Z_w = \sqrt{\mu_0/\epsilon_0}$ and $Z_s = \sqrt{j\omega\mu/\sigma}$, we have

$$R = 20 \log \frac{|Z_w|}{4|Z_s|} = 20 \log \left(\frac{1}{4}\left|\sqrt{\frac{\mu_0}{\epsilon_0}\frac{\sigma}{j\omega\mu}}\right|\right)$$

$$= 10 \log \left(\frac{\sigma_{Cu}}{16 \cdot 2\pi\epsilon_0}\right) + 10 \log \frac{\sigma_r}{f\mu_r}$$

$$= 108 - 10 \log \frac{f\mu_r}{\sigma_r} \text{ dB} \tag{9-48}$$

where f is in megahertz and $\sigma_{Cu} = 5.82 \cdot 10^7 \text{ U/m}$.

From Equation (9-46), we may establish a lower bound for M as follows:

$$M = 20 \log |1 - e^{-2t/\delta}e^{-j2t/\delta}| \geq 20 \log (1 - |e^{-2t/\delta}||e^{-j2t/\delta}|)$$

$$M \geq 20 \log (1 - e^{-2t/\delta}) = 20 \log (1 - 10^{-A/10}) \text{ dB} \tag{9-49}$$

Since M is a negative number it represents *gain*, that is, reduction of loss. This inequality assures that the loss will be no less than this lower bound.

It is best to plot A, R, and M on a semi-log grid. The nature of the functions is evident if we rewrite them using $\log f$ as the independent variable, as follows:

absorption loss

$$A = 132t\sqrt{\mu_r\sigma_r}(\sqrt{10})^{(\log f)} \text{ dB} \tag{9-50}$$

$$R = 108 + 10 \log \frac{\sigma_r}{\mu_r} - 10 \log f \text{ dB} \tag{9-51}$$

with f in megahertz and t in millimeters. Clearly, A is an exponential curve with respect to $\log f$, and R is a straight line. Their plots, for $t = 0.5$ mm and $\mu_r = \sigma_r = 1$, appear in Figure 9-5. As before, σ_r is with respect to annealed copper, for which $\sigma_{Cu} = 5.82 \cdot 10^7 \text{ U/m}$. Since the above analysis applies only to plane waves, it is important to note the minimum distance from the source at which it is usable. The minimum far-field distance for each frequency is given in Figure 9-5. At closer distances, we must use near-field techniques, to be discussed shortly. Obviously, at low frequencies, Figure 9-5 has little practical application. It will be useful, however, in the development of the near-field shielding effectiveness.

The preceding analysis of reflection loss considers only plane waves approaching a shield perpendicularly. For plane waves approaching obliquely, the analysis is slightly more complicated. We must consider the orientation of the **E** and **H** fields with respect to the shield surface. If the **E** field is parallel to the shield surface, the wave is *perpendicularly polarized* with respect to that surface. If the **H**

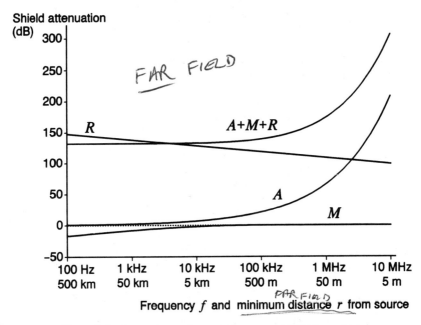

Figure 9-5   Attenuation vs. frequency for copper shield $1/2$ mm thick

field is parallel to the surface, the wave is *parallel polarized*. Examples of the two types of polarization appear in Figure 9-6. Their designations refer to the respective directions of the **E** field with respect to the plane of incidence, which is the plane of the printed page in Figure 9-6. The shield boundary conditions produce slightly different equations for the two cases. The most general case, where neither vector is parallel to the shield surface, may be split into one parallel and one perpendicularly polarized component. We will not study obliquely approaching waves here, but a complete discussion appears in the reference material.[2] Our earlier discussion of perpendicularly approaching waves provides a reasonable estimate of shielding effectiveness for all cases.

If the shield is in the near field of the approaching wave, the problem still consists of finding a solution to Maxwell's equations. The solution must satisfy the same boundary conditions at the surfaces of the shield that are given in Equations (9-23) and (9-24). As in the far field, the problem is solvable using wave and shield impedances. However, in the near field, although $Z_w = |\mathbf{E}|_s / |\mathbf{H}|_s$, Z_w is complex and is not simply equal to $\sqrt{\mu_0/\epsilon_0}$. It also depends on the distance r from the source. As shown in Chapter 8, near an electric source (for $\theta = \pi/2$),

$$Z_w = \frac{1}{j\omega r \epsilon} \tag{9-52}$$

2. Simon Ramo, John R. Whinnery, and Theodore Van Duzer, *Fields and Waves in Communication Electronics,* (New York, N.Y.: John Wiley & Sons, Inc., 1984), pp. 296–305.

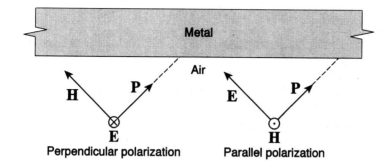

Figure 9-6 Polarizations of waves approaching shield obliquely.

and near a magnetic source,

$$Z_w = j\omega r\mu$$

affects reflection loss see page 144

(9-53)

Since near-field effects also may exist within the shield, we do not know that Z_s is equal to $\sqrt{j\omega\mu/\sigma}$. To evaluate Z_s rigorously would require an exact solution of Maxwell's equations at the shield surface. Instead, we will evaluate Z_s heuristically.

In the near or far-field regions, or anywhere between, the phases of all $\mathbf{E}$ and $\mathbf{H}$ elements depend on r but not θ or ϕ. For example, in a nonconducting medium (where $\alpha = 0$), if $\psi_{E\theta}$ is the phase of the θ component of $\mathbf{E}$, then

$$\psi_{E\theta} = \tan^{-1}\frac{\mathscr{I}\{E_\theta\}}{\mathscr{R}\{E_\theta\}} = -\beta r + \tan^{-1}\frac{\mathscr{I}\{e^{j\beta r}E_\theta\}}{\mathscr{R}\{e^{j\beta r}E_\theta\}}$$

$$= -\beta r + \tan^{-1}\frac{\dfrac{lQ}{4\pi\epsilon}\left(\dfrac{\beta}{r^2}\right)\sin\theta}{\dfrac{lQ}{4\pi\epsilon}\left(\dfrac{1}{r^3} - \dfrac{\beta^2}{r}\right)\sin\theta}$$

$$= -\beta r + \tan^{-1}\frac{r\beta}{1 - r^2\beta^2}$$

(9-54)

It may similarly be shown that the phases of E_r, E_ϕ, H_r, H_θ, and H_ϕ depend only on r, in any conducting or nonconducting medium. Therefore, on a spherical surface enclosing an electric or magnetic dipole source, the phases of all **E** and **H** components due to that source are constant. Their magnitudes, but not their phases, will vary with θ. The phase angle of Z_W or Z_s due to each field component is therefore also constant over such a surface.

Surfaces of constant phase, called equi-phase surfaces, are shown for an electric dipole source in Figure 9-7. The figure is not to scale, since, for example, $\lambda = 300$ m for a 1-MHz field in free space, whereas $\lambda = 0.655$ mm in an aluminum shield, as shown earlier. The equi-phase surfaces within the shield are in reality much flatter than shown, and are very nearly planar. Thus, even with a dipole source located only $\lambda/60$ from the shield surface, the field within the shield material behaves as a far field. Therefore, we calculate Z_s as before, and all other effects within the shield are the same.

The near field thus behaves like the far field except that Z_W is no longer equal to $\sqrt{\mu_0/\epsilon_0}$ but is instead given by Equation (9-52). This only affects the reflection loss R; the factors A and M remain the same. In the near field of an electric source,

$$R = 20 \log \frac{|Z_W|}{4|Z_s|} = 20 \log \frac{1/(j\omega r \epsilon_0)}{4|\sqrt{j\omega\mu/\sigma}|}$$

$$= 10 \log \frac{\sigma}{16\omega\mu(\omega r \epsilon_0)^2} = 10 \log \frac{\sigma}{16\omega^3 \mu r^2 \epsilon_0^2}$$

$$= 10 \log \frac{\sigma_{Cu}}{16 \cdot (2\pi)^3 \mu_0 \epsilon_0^2} + 10 \log \frac{\sigma_r}{f^3 \mu_r r^2}$$

$$= 142 + 10 \log \frac{\sigma_r}{f^3 \mu_r r^2} \text{ dB} \tag{9-55}$$

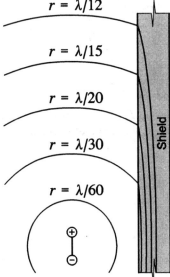

Figure 9-7 Equi-phase surfaces of wave penetrating shield (not to scale)

where f is in megahertz and r is in meters. Comparison with Equation (9-48) shows that for an electric field source the reflection loss is higher in the near field than the far field. If we know that the field source is electric, the actual loss can be no less than the far-field loss. It is therefore safe to assume that the far-field loss is the minimum attenuation provided by the shield.

In the near field of a magnetic source, Z_w is given by Equation (9-53), and R becomes

$$R = 20 \log \frac{|Z_w|}{4|Z_s|} = 20 \log \frac{|j\omega r \mu_0|}{4|\sqrt{j\omega\mu/\sigma}|}$$

$$= 10 \log \frac{\sigma(\omega r \mu_0)^2}{16\omega\mu} = 10 \log \frac{\sigma\omega r^2 \mu_0}{16\mu_r}$$

$$= 10 \log \frac{2\pi\sigma_{Cu}\mu_0}{16} + 10 \log \frac{\sigma_r f r^2}{\mu_r}$$

$$= 74.6 + 10 \log \frac{\sigma_r f r^2}{\mu_r} \text{ dB} \qquad (9\text{-}56)$$

where, as before, f is in megahertz and r is in meters. Comparison with Equation (9-48) shows that for a magnetic source the loss is lower in the near field than the far field. If the field source is magnetic, then this near-field loss is the only safe value to assume for the shield attenuation. We also must assume this value if we do not know the nature of the source.

The reflection loss is always a function of σ_r/μ_r and is lowest for low-frequency magnetic sources. Equations (9-48), (9-55), and (9-56) give the asymptotic values for the reflection loss. The exact values, obtained by setting $Z_w = |E|_s/|H|_s$ and substituting it into Equation (9-44), are plotted in Figure 9-8. Above 10 MHz, any practical shield *with no holes* provides at least 200 dB attenuation, even for the field near a magnetic source. From Figure 9-8 it is obvious that the most difficult field to shield is a low-frequency magnetic field. The reason is that both the absorption and reflection losses are low, and the gain due to internal reflections is not negligible. Due to the approximations assumed in the formulas, the calculations might even show a net gain instead of a loss. Of course this represents no actual gain, but it does mean that the shield is ineffective at those frequencies. The absorption loss can be increased by using a material with a higher $\sigma\mu$ product.

9-3 DIVERSION OF LOW-FREQUENCY MAGNETIC FIELDS BY HIGH-PERMEABILITY SHIELDS

We noted in the preceding section that in the near field resulting from a magnetic source there is no effective method of reducing the field. In these cases, however, it is usually possible to *divert* the magnetic field using a high-permeability material. This is a completely different phenomenon and does not depend on conductivity or

non - magnetic

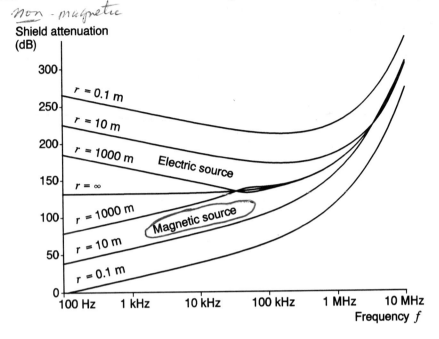

Shield attenuation
(dB)

Figure 9-8   Near- and far-field attenuation

frequency (except that the permeability may vary with frequency). It is typically used around magnetic recording heads and some transformers.

In the near field of a magnetic source, the electric field becomes negligible. Then we need not consider a traveling electromagnetic wave, but only a stationary magnetic field. The field may truly be stationary due to a permanent magnet or a steady electric current. More commonly, the field is changing, but slowly enough so that it behaves as if it were stationary.

To analyze this situation, we must consider a completely closed shield. We immerse the enclosure in a uniform magnetic field and find the intensity of the magnetic field inside the enclosure. Any discontinuities in the shield, such as edges and corners, cause complications in the field expressions. In the earlier cases of traveling waves, we avoided these discontinuities by restricting our discussion to a shield consisting of an infinite flat surface. Since we now require a closed surface, we will approximate the actual shield by using a hollow sphere. This permits an exact analytical solution. The field due to a cubic or solid rectangular shield is much more complicated, but the spherical approximation is adequate for most purposes.

At the frequencies of concern, the magnetic field is effectively unrelated to the electric field. Since the magnetic shield does not attenuate the field but diverts it, the normal approach is to consider the magnetic flux density **B**. The unit of magnetic flux density is the *tesla*, abbreviated *T,* though some older literature uses the *gauss*. One tesla is equal to 1 Wb/m^2, and one gauss is 10^{-4} T.

Consider a hollow spherical shield of relative permeability μ_r immersed in a magnetic field that was uniform before the sphere was introduced. The scheme appears in Figure 9-9. The shield affects the nearby field, but the distant field is still uniform. Let $B_0\mathbf{k}$ be the distant field, where $\mathbf{k}$ is a unit vector in the z direction. Let a and b be the inner and outer radii of the sphere, respectively.

To find an exact expression for $\mathbf{B}$, we must solve Maxwell's divergence equation for magnetic fields:

$$\nabla \cdot \mathbf{B} = 0 \qquad (9\text{-}57)$$

His other three equations become trivial since they involve electric fields, which we now assume to be zero. There are three distinct regions: inside the hollow space, within the shield material itself, and outside the shield. There can be a different solution to Equation (9-57) for each region. All three solutions must also satisfy boundary conditions at the surfaces between the regions.

Since the magnetic field is not due to any electric current near the shield, the magnetic field is conservative. We may therefore use the magnetic scalar potential U to simplify the analysis, where

$$\nabla U = -\mathbf{H} = -\mathbf{B}/\mu \qquad (9\text{-}58)$$

With this definition, solving Equation (9-57) reduces to solving Laplace's equation:

$$\nabla^2 U = \nabla \cdot (\nabla U) = -\nabla \cdot \mathbf{B}/\mu = 0 \qquad (9\text{-}59)$$

Due to the spherical symmetry, we again use a spherical coordinate system, this time with its origin at the center of the shield. Laplace's equation may be solved

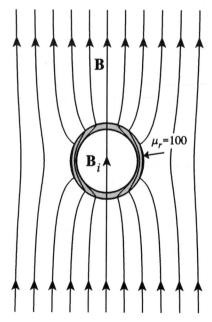

Figure 9-9 Deflection of magnetic field by high-permeability shield

using a product solution.[3] For the static case considered here, the most general solution of interest is

$$U = (C_1 r + C_2 r^{-2}) \cos\theta \qquad (9\text{-}60)$$

Although there are higher-order Legendre polynomials that also satisfy Laplace's equation, they become zero for the static case.

    At large distances from the shield, the magnetic flux density must be uniform and equal to $B_0 \mathbf{k}$. For large r, Equation (9-60) becomes

$$U = C_1 r \cos\theta = C_1 z \qquad (9\text{-}61)$$

The magnetic flux density is then

$$\mathbf{B} = -\mu\nabla U = -\mu\nabla C_1 z = -\mu C_1 \mathbf{k} \qquad (9\text{-}62)$$

Since this flux density must be equal to $B_0 \mathbf{k}$, the constant C_1 in Equation (9-60) must be equal to $-B_0/\mu_0$ in the region outside the shield. We will designate the other constant C_2 as C_o for this region. The potential U_o everywhere outside the shield is therefore

$$U_o = (-B_0 r/\mu_0 + C_o r^{-2}) \cos\theta \qquad (9\text{-}63)$$

    At the origin, inside the hollow portion of the shield, the potential must remain finite, since there is no magnetic source inside the shield. This is possible only if the constant C_2 in Equation (9-60) is zero for this region. With the other constant C_1 designated as C_i for the hollow region, the potential U_i in this region is

$$U_i = C_i r \cos\theta \qquad (9\text{-}64)$$

    In the region between the shield surfaces, within the metal of the shield, both constants have values that depend on the boundary conditions. We will designate them as C_{s1} and C_{s2} for this region. The potential U_s within the metal of the shield is then

$$U_s = (C_{s1} r + C_{s2} r^{-2}) \cos\theta \qquad (9\text{-}65)$$

The problem that remains is to find values for C_o, C_{s1}, C_{s2}, and C_i. To find these four unknowns requires four boundary conditions, which we will now formulate.

    Since there are no electrical currents near the shield, the magnetic field is conservative. Therefore, at the outer surface of the shield, the potential outside the shield must be equal to the potential within the metal. Symbolically,

$$U_o \big|_{r=b} = U_s \big|_{r=b}$$

$$(-B_0 b/\mu_0 + C_o b^{-2}) \cos\theta = (C_{s1} b + C_{s2} b^{-2}) \cos\theta$$

$$-B_0 b/\mu_0 + C_o b^{-2} = C_{s1} b + C_{s2} b^{-2} \qquad (9\text{-}66)$$

---

3. Ibid., pp. 378–80.

Similarly, the potentials at the inner surface must be equal:

$$U_i \big|_{r=a} = U_s \big|_{r=a}$$

$$C_i a \cos\theta = (C_{s1}a + C_{s2}a^{-2}) \cos\theta$$

$$C_i a = C_{s1}a + C_{s2}a^{-2} \tag{9-67}$$

These form two of the required boundary conditions. The others relate to continuity of magnetic flux across the boundaries. By Gauss's law for magnetic flux, the normal components of **B** must be equal on both sides of a boundary surface. At each shield surface, the normal component of **B** is the radial component. In the three regions, we designate these components $(B_o)_r$, $(B_s)_r$, and $(B_i)_r$, respectively. From the definition of U in Equation (9-58), we may express the continuity at the outer surface as

$$(B_o)_r \big|_{r=b} = (B_s)_r \big|_{r=b}$$

$$\mu_0 \frac{\partial U_o}{\partial r} \bigg|_{r=b} = \mu_r \mu_0 \frac{\partial U_s}{\partial r} \bigg|_{r=b}$$

$$\mu_0(-B_0/\mu_0 - 2C_o b^{-3}) \cos\theta = \mu_r \mu_0 (C_{s1} - 2C_{s2}b^{-3}) \cos\theta$$

$$-B_0/\mu_0 - 2C_o b^{-3} = \mu_r C_{s1} - 2\mu_r C_{s2}b^{-3} \tag{9-68}$$

Similarly, at the inner surface,

$$(B_i)_r \big|_{r=a} = (B_s)_r \big|_{r=a}$$

$$\mu_0 \frac{\partial U_i}{\partial r} \bigg|_{r=a} = \mu_r \mu_0 \frac{\partial U_s}{\partial r} \bigg|_{r=a}$$

$$\mu_0 C_i \cos\theta = \mu_r \mu_0 (C_{s1} - 2C_{s2}a^{-3}) \cos\theta$$

$$C_i = \mu_r C_{s1} - 2\mu_r C_{s2}a^{-3} \tag{9-69}$$

Equations (9-66) through (9-69) are four linear equations in the four unknowns C_o, C_{s1}, C_{s2}, and C_i. We may solve for all of them and substitute into Equations (9-63) through (9-65) to obtain the magnetic potentials in the three regions, and from them obtain the flux densities. The flux lines in Figure 9-9 were obtained in this manner, for $\mu_r = 100$ and $b = 1.2a$. This is not a very practical shield, but its dimensions allow the reader to visualize the process. Normally, the shield would be thinner and μ_r would be much higher. Values of 20,000 or higher are possible, using mumetal or other alloys designed for this purpose.

Since we are normally concerned only with the flux in the hollow space inside the shield, we need only solve for C_i. Elimination of the other variables yields

$$C_i = \frac{-B_0/\mu_0}{1 - \dfrac{2(\mu_r + 1)^2}{9\mu_r}\left[1 - \left(\dfrac{a}{b}\right)^3\right]} \tag{9-70}$$

Substituting C_i into Equation (9-64) and then using Equation (9-58) gives the flux density $\mathbf{B}_i$ in the hollow space:

$$\mathbf{B}_i = -\mu_0 \nabla U_i = -\mu_0 C_i \mathbf{k} = \frac{B_0 \mathbf{k}}{1 + \dfrac{2(\mu_r + 1)^2}{9\mu_r}\left[1 - \left(\dfrac{a}{b}\right)^3\right]} \tag{9-71}$$

The field inside the shield is therefore uniform and is in the same direction as the original exterior field. The shield attenuation is the ratio of these fields, or

$$A = 20 \log \frac{B_0}{|\mathbf{B}_i|} = 20 \log \left\{ 1 + \frac{2(\mu_r + 1)^2}{9\mu_r}\left[1 - \left(\frac{a}{b}\right)^3\right] \right\} \tag{9-72}$$

For the shield of Figure 9-9, the attenuation is approximately 20 dB, or a factor of 10. Usually the thickness of the shield is much less than its radius, and then we let $t = b - a$. The attenuation is then approximately

$$A = 20 \log \left\{ 1 + \frac{2(\mu_r - 1)^2}{9\mu_r}\left[1 - \left(1 - \frac{t}{b}\right)^3\right] \right\}$$

$$\approx 20 \log \left\{ 1 + \frac{2(\mu_r - 1)^2}{9\mu_r}\left[1 - \left(1 - 3\frac{t}{b}\right)\right] \right\}$$

$$= 20 \log \left[1 + \frac{2(\mu_r - 1)^2 t}{3\mu_r b} \right] \tag{9-73}$$

For a mumetal shield ($\mu_r = 20{,}000$) with $t = 0.01b$, the attenuation is 42.4 dB, or a factor of 131. Unlike absorption loss, attenuation due to diversion of flux is *not* directly proportional to the shield thickness.

The above analysis is applicable only at very low frequencies. The relative permeability μ_r of mumetal is very high, typically 20,000 at these frequencies. As the frequency increases, however, the permeability of mumetal and most other ferromagnetic materials decreases very rapidly. Fortunately, at higher frequencies, absorption loss begins to take effect, and magnetic field diversion is less necessary. Absorption loss, however, depends on the $\mu_r \sigma_r$ *product*. The relative conductivity σ_r is only 0.03 for mumetal but is 0.1 for steel. At frequencies between approximately 10 kHz and 1 MHz, it can be shown that $\mu_r \sigma_r$ is highest for steel. At these frequencies, steel is usually the best ordinary shielding material. At still higher frequencies, μ_r becomes less than $1/\sigma_r$ for steel and mumetal, so that $\mu_r \sigma_r < 1$. Then copper, for which $\mu_r \sigma_r = 1$, has the highest $\mu_r \sigma_r$ product and thus provides more absorption.

Absorption and magnetic diversion are thus effective over different frequency ranges. With proper choice of shielding materials, the frequency ranges will overlap, and the materials will provide effective shielding at all frequencies. The designer must make sure, however, that there is no intermediate frequency range where neither absorption nor magnetic diversion provides sufficient attenuation. If there is, a thicker shield is usually necessary.

It is also important to consider the saturation effects of ferromagnetic materials, since their permeability also decreases in the presence of strong magnetic fields. Unfortunately, materials of higher permeability also saturate more quickly. We obtain the best performance by using a multilayer laminate of materials having different permeabilities, with the lowest on the side nearest the source. The strongest field then irradiates the layer of lowest permeability. This material does not saturate as much as the higher-permeability material would saturate if exposed to the same field. Then, when the weakened field reaches the higher-permeability layer, that material saturates less, and it therefore attenuates the remaining field by a greater factor. For highest attenuation, the layers should be separated by air gaps at least as thick as the shield layers, to avoid interaction of fields in adjacent layers. Also, intense direct currents should not be allowed to flow through or near a magnetic shield of high permeability.

9-4 PROBLEMS

1. What is the skin depth for annealed copper at a frequency of 10 MHz? At 1000 MHz? How thick must a copper shield be to provide an absorption loss of 100 dB at each of these frequencies?

2. The relative conductivity of a certain grade of steel is 0.1, and its relative permeability is 500 at a frequency of 10 MHz. Compute the skin depth for this material at that frequency. How thick must a steel shield be to provide an absorption loss of 100 dB?

3. From Figure 9-2 it is evident that $\mathbf{E}_r$ points in the $-\mathbf{a}_E$ direction and $\mathbf{H}_r$ points in the $+\mathbf{a}_H$ direction, where $\mathbf{a}_E$ and $\mathbf{a}_H$ are the directions of $\mathbf{E}_0$ and $\mathbf{H}_0$, respectively.
   (a) Show that

$$(\mathbf{E}_r)_s = -E_r e^{j\beta r}\, \mathbf{a}_E \qquad (9\text{-}74)$$

$$(\mathbf{H}_r)_r = \frac{E_r}{\eta_0} e^{j\beta r}\, \mathbf{a}_H \qquad (9\text{-}75)$$

   is a solution to Maxwell's equations satisfying this boundary condition.
   (b) How do we know that this wave is traveling in the $-r$ direction?
   (c) Show that

$$(\mathbf{E}_r)_s = -E_r e^{-j\beta r}\, \mathbf{a}_E \qquad (9\text{-}76)$$

$$(\mathbf{H}_r)_s = \frac{E_r}{\eta_0} e^{-j\beta r}\, \mathbf{a}_H \qquad (9\text{-}77)$$

   is *not* a solution.

4. What is the far-field reflection loss at the surface of an annealed copper shield at the frequencies of 10 MHz and 1000 MHz?

5. What is the far-field reflection loss at the surface of a steel shield at the frequencies of 10 MHz and 1000 MHz? The relative permeability of the steel used in the shield is 500 at 10 MHz and 50 at 1000 MHz. The relative conductivity is 0.1 at all frequencies.

6. Compute the total attenuation of a 100-MHz far-field wave due to a steel shield that is 0.1 mm thick. The relative permeability is 100 at that frequency.

7. Repeat Problem 6 for a copper shield of the same thickness.

8. The relative permeability of a certain steel shield is 700 at 1 MHz and 1000 at 60 Hz, and the shield is 0.1 mm thick.
   (a) Compute the total attenuation of this shield when located 0.1 m away from a 1-MHz magnetic-field source.
   (b) Repeat for a 60-Hz magnetic-field source at the same distance.

9. A personal computer is enclosed in a molded plastic cabinet with a uniform wall thickness of 3 mm. Its radiated emissions at 150 MHz are 30 dB too high. The designer proposes to make the plastic material conductive, to act as a shield. The cabinet is 0.1 m away from the sources, most of which are current loops. What must be the conductivity of the plastic (for which $\mu_r = 1$) to provide the required 30 dB of attenuation?

10. With the approximations used to derive Equation (9-44), it appears that as the shield thickness t approaches zero, A approaches zero, R remains constant, and M approaches $-\infty$. By using exact values for R and Γ, show that, instead

$$\lim_{t \to 0} M = -R \qquad\qquad \textbf{(9-78)}$$

and thus show that the shielding effectiveness decreases smoothly to zero as thickness decreases.

11. Show that Equation (9-60) is a solution to Laplace's equation. The Laplacian operation in spherical coordinates is

$$\nabla^2 U = \frac{1}{r^2} \cdot \frac{\partial}{\partial r}\left(r^2 \frac{\partial U}{\partial r}\right) + \frac{1}{r^2 \sin\theta} \cdot \frac{\partial}{\partial \theta}\left(\sin\theta \frac{\partial U}{\partial \theta}\right) + \frac{1}{r^2 \sin^2\theta} \cdot \frac{\partial^2 U}{\partial \phi^2} \qquad \textbf{(9-79)}$$

12. An enclosed cubical steel shield is 10 cm on each edge, and is made of steel 1 mm thick. By approximating the cube with an inscribed sphere, estimate the amount by which the shield attenuates a low-frequency external magnetic field. The relative permeability of the shield is 1000.

13. Repeat Problem 12 for a mumetal shield of the same dimensions, for which the relative permeability is 20,000.

10

EFFECTS OF
SHIELD APERTURES

In the analyses performed in Chapter 9 we assumed all shields to be solid. Of course, this is unrealistic, and we must investigate the effects of holes in the shield, which cause leakage and must be analyzed as slot antennas. Even worse is a hole with a wire or cable passing through it, which can render a shield useless unless corrective measures are taken. These will be discussed in Chapter 11. A hole with no conductor passing through it does not greatly reduce the shielding effectiveness if the hole is small enough. We shall now study the factors that limit the acceptable size of a hole in a shield.

10-1 CURRENT FLOW IN SHIELDS

A conductive shield performs its function by allowing current to flow in it. Ampere's law, which is one of Maxwell's curl equations, determines this current. When we solved Maxwell's equations for a conductive material in Chapter 9, we implicitly defined this current. We noted that this current, $\mathbf{J}$, is equal to $\sigma \mathbf{E}$. We will now study the current in more detail.

Ampere's law in phasor form is

$$\nabla \times \mathbf{H}_s = \mathbf{J}_s + j\omega \mathbf{D}_s = (\sigma + j\omega\epsilon)\, \mathbf{E}_s \qquad (10\text{-}1)$$

We again assume the shield to be a good conductor, for which $\sigma \gg j\omega\epsilon$. Within the shield, Equation (10-1) then becomes

$$\mathbf{J}_s = \sigma \mathbf{E}_s = \nabla \times \mathbf{H}_s \qquad (10\text{-}2)$$

We know $\mathbf{E}$ and $\mathbf{H}$ as a function of depth into the shield, and we may use either to calculate $\mathbf{J}$ in Equation (10-2). Although it is simpler to use $\sigma \mathbf{E}$, the physical interpretation of $\nabla \times \mathbf{H}$ is more important.

Consider a far-field wave approaching a shield perpendicularly. Let $\mathbf{E}_0$ and $\mathbf{H}_0$ be the incident fields, $\mathbf{E}_r$ and $\mathbf{H}_r$ the reflected fields, and $\mathbf{E}_1$ and $\mathbf{H}_1$ the fields just beyond the nearer shield surface, as in Chapter 9. For simplicity, we consider a shield thick enough that the fields are negligible after penetrating the thickness of the shield. Then there are no internal reflections, and $\mathbf{E}_1$ and $\mathbf{H}_1$ are the only fields in the shield material adjacent to its nearer surface. If $\mathbf{J}$ is the current in the shield, then its time magnitude is

$$|\mathbf{J}_s| = \sigma|\mathbf{E}_s| = \sigma|(\mathbf{E}_1)_s|e^{-\alpha r} \tag{10-3}$$

Therefore, the current $\mathbf{J}$, like the fields $\mathbf{E}$ and $\mathbf{H}$, decreases exponentially with depth into the shield. The current $\mathbf{J}$ and the internal field $\mathbf{H}$ appear in Figure 10-1, along with the incident and reflected fields $\mathbf{E}_0$, $\mathbf{H}_0$, $\mathbf{E}_r$, and $\mathbf{H}_r$. Because the internal field $\mathbf{E}$ is very small, it does not appear in the figure, but it is proportional to $\mathbf{J}$ and in the same direction.

The integral of the current density $\mathbf{J}$ through the thickness of the shield is the sheet current density $\mathbf{K}$ in the shield. For a shield of the thickness considered here, $e^{-\gamma t} \ll 1$, and therefore

$$\mathbf{K}_s = \int_0^t \mathbf{J}_s dr = \int_0^t \sigma(\mathbf{E}_1)_s e^{-\gamma r} dr = -\sigma \frac{(\mathbf{E}_1)_s}{\gamma} e^{-\gamma r}\Big|_0^t$$

$$= \frac{\delta\sigma(\mathbf{E}_1)_s}{1+j}(1 - e^{-\gamma t}) \approx \frac{\delta\sigma(\mathbf{E}_1)_s}{1+j} = \frac{\delta}{1+j}(\mathbf{J}_1)_s \tag{10-4}$$

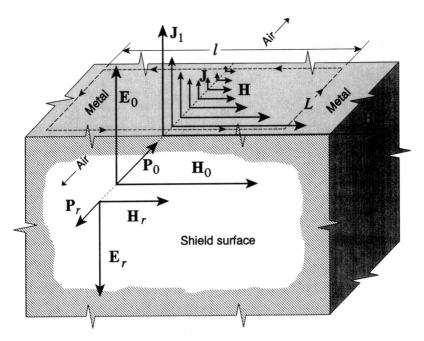

Figure 10-1 Magnetic fields and currents within a metallic shield

where $\mathbf{J}_1$ is the current density at the shield surface. If we integrate $\mathbf{H}$ around the rectangular path L shown by the dotted line in Figure 10-1, the only significant contribution to the integral is from the field $\mathbf{H}_1$ on the near side of the shield. If l is the length of the rectangle and S is its area, then by Ampere's law,

$$\oint_L \mathbf{H}_s \cdot d\mathbf{L} = |\mathbf{H}_1|_s\, l = \int_S \mathbf{J}_s \cdot d\mathbf{S} = |\mathbf{K}|_s\, l$$
$$|\mathbf{H}_1|_s = |\mathbf{K}|_s \qquad\qquad (10\text{-}5)$$

The magnitude of $\mathbf{K}$ is thus equal to $\mathbf{H}_1$, and its direction is that of $\mathbf{E}_0$. That direction depends on the characteristics of the source or sources, which are often unknown. The fields might not propagate directly from a source to the shield, but instead may be reflected at one or more intermediate surfaces. They may even be circularly or elliptically polarized, and thus contain two out-of-phase components. Fields at different frequencies often are oriented in different directions. The direction of $\mathbf{J}$ is therefore very difficult to predict and may change due to the slightest change in geometry.

Any holes in the shield divert the current. Consider holes in a shield of the shapes shown in Figures 10-2 and 10-3. For current flowing vertically, as shown, the slot diverts nearly as much current as the round hole. The two apertures would cause about equal degradation of the shield, for the current shown.

The change in current flow, due to a hole in the shield, may be regarded as a new current $\mathbf{J}_a$. This new current adds to the original current $\mathbf{J}$ and generates

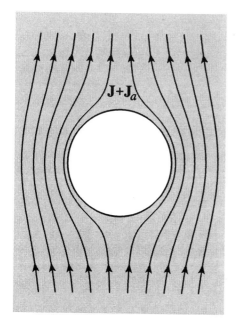

Figure 10-2 Shield currents deflected by a round hole

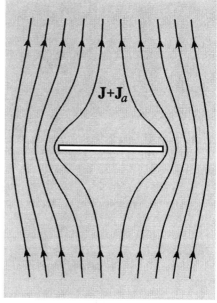

Figure 10-3 Shield currents deflected by a narrow slot

additional fields. The total current $\mathbf{J} + \mathbf{J}_a$ must satisfy the continuity equation $\nabla \cdot (\mathbf{J} + \mathbf{J}_a) = 0$ within the shield material, and in the hole, of course, $\mathbf{J} + \mathbf{J}_a = 0$. An exact expression for $\mathbf{J}_a$ is possible only for simple shapes. For a circular hole in a shield that is *thin* compared to the skin depth, however, a simple expression for $\mathbf{J}_a$ results. This is not a normal situation, since the absorption loss is then negligible. It nevertheless allows us to plot the streamlines of $\mathbf{J}_a$ and obtain a qualitative understanding of more complex geometries.

For a shield that is thin compared to δ, there is no variation in the direction perpendicular to the shield surface, and we may use cylindrical coordinates. Let a be the radius of the hole. The current in the shield at a large distance from the hole is uniform and is equal to $\mathbf{J}$. Let the coordinate system be oriented so that $\mathbf{J}$ is in the y direction, or

$$\mathbf{J} = J_0\mathbf{j} = J_0(\sin\phi \, \mathbf{a}_\rho + \cos\phi \, \mathbf{a}_\phi) \tag{10-6}$$

where $J_0 = |\mathbf{J}|_s$, the spatial magnitude of $\mathbf{J}$. The total current $\mathbf{J} + \mathbf{J}_a$ must approach this value as ρ becomes large, which is one boundary condition. The other condition is that the radial component of $\mathbf{J} + \mathbf{J}_a$ must be zero at $\rho = a$, since no current can flow into or out of the hole. For a hole that is small compared to wavelength, we may treat the field as static. Then $\mathbf{J} + \mathbf{J}_a = \sigma\mathbf{E} = -\sigma\nabla V$ and the continuity equation becomes Laplace's equation. The solution that satisfies the above boundary conditions is

$$\mathbf{J} + \mathbf{J}_a = J_0\left(1 - \frac{b^2}{r^2}\right)\sin\phi \, \mathbf{a}_\rho + J_0\left(1 + \frac{b^2}{r^2}\right)\cos\phi \, \mathbf{a}_\phi \tag{10-7}$$

The difference between Equations (10-6) and (10-7) gives $\mathbf{J}_a$ within the shield:

$$\mathbf{J}_a = J_0\frac{b^2}{r^2}(-\sin\phi \, \mathbf{a}_\rho + \cos\phi \, \mathbf{a}_\phi) \tag{10-8}$$

Across the hole, the two currents must cancel, and

$$\mathbf{J}_a = -\mathbf{J} \tag{10-9}$$

The current $\mathbf{J}_a$ flows locally near the hole, and we consider it to flow across the hole, as shown in Figure 10-4. When the two currents are added, the total current resembles Figure 10-2.

For more complex shapes such as a rectangular hole, or for thicker shields where the current varies in the z direction, an exact solution is impractical. We may still use the above analysis to visualize the current $\mathbf{J}_a$ qualitatively. We consider $\mathbf{J}$ to be uniform over the entire shield, including the hole. This current causes the shield to behave as if it were solid. Then we consider $\mathbf{J}_a$ separately as a new source. We pretend that this current generates the $\mathbf{H}$ field that in reality passes through the hole.

There is also an $\mathbf{E}$ field passing through the hole, but we cannot generate it at the hole using a fictitious electric current. However, if we extend our imagination and invent a *magnetic current* whose density is $\mathbf{M}$, then we can use this current to

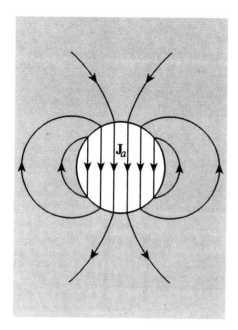

Figure 10-4 Additional current caused by
aperture in shield

generate **E** at the hole. With this new concept, Maxwell's second curl equation
(Faraday's law) becomes, in phasor form

$$\nabla \times \mathbf{E} = -\mathbf{M} - j\omega \mathbf{B} \qquad (10\text{-}10)$$

To generate the **E** field in the proper direction, **M** must therefore point in the same
direction as **H**. The two fictitious currents $\mathbf{J}_a$ and **M** together generate electromagnetic fields that emulate the actual fields passing through the hole.

The shape of hole that usually causes the most difficulty is a long, narrow slot.
A joint between two panels may appear "airtight," but even if the panels overlap,
the slot formed by the joint may cause radiation leakage. Therefore, we will now
study in detail the case of a narrow slot.

10-2 SLOT ANTENNA THEORY[1]

It might appear that the radiated fields passing through a slot would decrease to zero
as the slot width becomes smaller. Unfortunately, this is true only for a perfectly
absorbing shield, which is not realizable. To reduce the fields to zero using a
nonabsorbing shield, the capacitance across the slot must become infinite, which is
also not practical.

---

1. Edward C. Jordan and Keith G. Balmain, *Electromagnetic Waves and Radiating Systems,* 2nd
ed. (Englewood Cliffs, N.J.: Prentice Hall Press, 1968), pp. 513–19. Adapted by permission of Prentice
Hall Press, Englewood Cliffs, New Jersey.

A slot has the greatest effect when the shield current is perpendicular to it, as shown in Figure 10-3. The current $\mathbf{J}_a$ appears as current sources connected across the slot, as shown in Figure 10-5. The inductive impedance in the shield material, and the capacitive admittance across the slot, cause a time-varying potential to appear across the slot. If we knew the current, impedances, and admittances, as distributed parameters, we could calculate the potential. We could then calculate the field radiated from the slot by considering the slot as a multitude of oscillating dipoles.

The current sources are all equal, and their total current is $|\mathbf{K}|l$, where l is the slot length and $\mathbf{K}$ is the sheet current density in the shield, equal in magnitude to $|\mathbf{H}_1|$. The impedances and admittances, however, are more complicated. They are not simply the distributed inductance and capacitance, because the radiated field adds a resistive component to each quantity. To calculate them, we must solve Maxwell's equations subject to the boundary conditions of the slot, which leads to great difficulties. Instead, we will associate the problem with a more familiar one so that we can use solutions found in the literature. The circuit shown in Figure 10-5, however, provides a useful qualitative understanding of the process that occurs at a slot in a shield.

Since the current sources inject the current $\mathbf{J}_a$ at the edges of the slot, $\mathbf{J}_a$ flows on *both* sides of the shield, though the shield may be much thicker than the skin depth. The potential difference, and the resulting electric fields, appear as shown in Figure 10-6. The shield can have no tangential $\mathbf{E}$ field component at its surface, so the $\mathbf{E}$ field must be perpendicular to the shield over its surface. The $\mathbf{E}$ fields are symmetric on the two sides of the shield, so the $\mathbf{E}$ field within the slot must lie in the plane of the shield. These boundary conditions uniquely determine the $\mathbf{E}$ field, which in turn determines the $\mathbf{H}$ field.

On a dipole antenna of the same width as the above slot, the fields would be as shown in Figure 10-7. Since a conductor cannot have a time-varying normal $\mathbf{H}$ component, the $\mathbf{H}$ field generated by the dipole must be tangential to the antenna surface. The $\mathbf{H}$ field is symmetric about the plane containing the dipole antenna

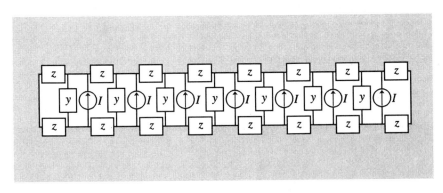

Figure 10-5   Aperture current modeled as current sources

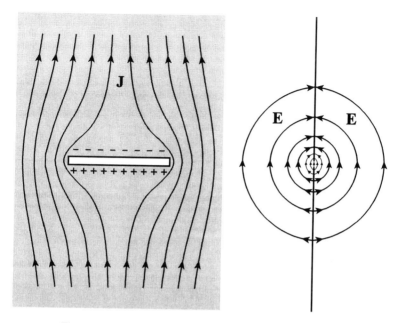

Figure 10-6   Electric fields near shield slot (orthographic view)

and must therefore intersect the plane perpendicularly. These boundary conditions uniquely determine the **H** field and in turn the **E** field.

If we consider only one side of the plane at a time, the slot and the dipole are *duals*. The fields are identical except that **E** and **H** are interchanged and the direction of one of them is reversed. We analyze the slot simply by interchanging the **E** and **H** fields, reversing one field, and treating the slot as a dipole antenna. More precisely, the **E** field of the slot is *proportional* to the **H** field of the dipole, and vice

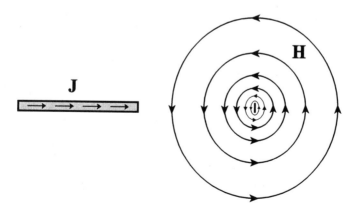

Figure 10-7   Magnetic fields near dipole (orthographic view)

versa. If $\mathbf{E}_a$ and $\mathbf{H}_a$ are the far fields of the slot and $\mathbf{E}_d$ and $\mathbf{H}_d$ are the far fields of the dipole, then

$$\mathbf{E}_a = k_1\mathbf{H}_d \qquad \mathbf{H}_a = k_2\mathbf{E}_d \qquad\qquad \textbf{(10-11)}$$

We must find the scale factors k_1 and k_2.

The ratio of $|\mathbf{E}|$ to $|\mathbf{H}|$ at large distances from the slot or dipole must be η_0, or 377 ohms, for both the slot and the dipole. Therefore, the magnitudes of k_1 and k_2 must be related as follows:

$$\eta_0 = \frac{|\mathbf{E}_a|}{|\mathbf{H}_a|} = \frac{|k_1\mathbf{H}_d|}{|k_2\mathbf{E}_d|} = \frac{|k_1|}{|k_2|\,\eta_0}$$

$$\left|\frac{k_1}{k_2}\right| = \eta_0^2 \qquad\qquad \textbf{(10-12)}$$

The waves propagate in the same direction for the two cases. It is convenient to scale the fields so that the respective power densities are equal. Therefore,

$$\mathbf{E}_d \times \mathbf{H}_d = \mathbf{E}_a \times \mathbf{H}_a = (k_1\mathbf{H}_d) \times (k_2\mathbf{E}_d) = -k_1k_2\mathbf{E}_d \times \mathbf{H}_d$$

$$k_1k_2 = -1 \qquad\qquad \textbf{(10-13)}$$

Therefore, if we choose the signs so that $\mathbf{E}_a$ and $\mathbf{H}_d$ are in the same direction,

$$k_1 = \eta_0 \qquad k_2 = -1/\eta_0 \qquad\qquad \textbf{(10-14)}$$

Thus,

$$\mathbf{E}_a = \eta_0\mathbf{H}_d \qquad \mathbf{E}_d = -\eta_0\mathbf{H}_a \qquad\qquad \textbf{(10-15)}$$

A completely free-floating dipole antenna, when met by an electromagnetic wave, will absorb some energy and then, as a new source, reradiate that energy. A slot in an infinitely large shield will do the same, with the $\mathbf{E}$ and $\mathbf{H}$ fields interchanged and one sign reversed. The slot behaves as the circuit of Figure 10-5, and the equivalent circuit for the dipole is shown in Figure 10-8. The voltage sources are due to the time-varying $\mathbf{H}$ field intersecting the dipole, and they represent the received signal voltage. If we replace the electric conductor of the dipole with a fictitious magnetic conductor, the fields on one side of it will be identical with those of the slot. On the other side, both the $\mathbf{E}$ and $\mathbf{H}$ fields will be opposite to those of

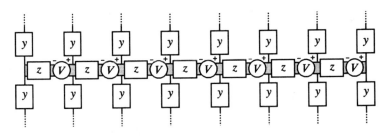

Figure 10-8   Dipole antenna modeled as voltage sources

the slot. They would be equal to the fields of the slot if we reversed the direction of the magnetic current. If we can find the fields for the dipole, we can easily calculate the fields for the slot using Equation (10-15).

Since a dipole antenna is probably more familiar than a slot, we will study the dipole first and then interchange the two fields. To find the fields of the dipole, we must know the current in it at all points. Each voltage source in Figure 10-8 generates current all along the dipole, just as each current source in Figure 10-5 generates voltage all along the slot. For the dipole, the ratio of the induced current at point x_1 to the source voltage at point x_2 is the transfer admittance between those points. We will first obtain a general expression for the transfer admittance $y(x_1, x_2)$ between any two points on the dipole. From this we will find the current at x_1 by integrating over all voltage sources. We can then find the total radiated field by integrating the current over the length of the dipole.

By reciprocity, $y(x_1, x_2) = y(x_2, x_1)$. To find $y(x_2, x_1)$, we set all voltage sources to zero except the one at x_1, which we call V_1. Let $I_1(x_2)$ be the current at x_2 due to the source at x_1. Then

$$y(x_2, x_1) = \frac{I_1(x_2)}{V_1} \tag{10-16}$$

The current $I_1(x_2)$ on each of the two sections of a dipole antenna driven at a single point is approximately sinusoidal[2]. A slope discontinuity occurs at the driven point. The distance between the zeroes of each sine function is approximately equal to $\lambda/2$, or half the free-space wavelength of the source frequency. Mathematically, if the origin is at the center of the dipole,

$$I_1(x_2) = \begin{cases} \dfrac{I_1(x_1) \sin[\beta(l/2 + x_2)]}{\sin[\beta(l/2 + x_1)]} & \text{for } -\dfrac{l}{2} \le x_2 \le x_1 \\[3mm] \dfrac{I_1(x_1) \sin[\beta(l/2 - x_2)]}{\sin[\beta(l/2 - x_1)]} & \text{for } x_1 \le x_2 \le \dfrac{l}{2} \end{cases} \tag{10-17}$$

where $\beta = 2\pi/\lambda$ as usual. We now must find $I_1(x_1)$.

The ratio $V_1/I_1(x_1)$ is the driving-point impedance $Z(x_1)$ of the antenna at the point x_1. At most points it is primarily reactive, and its reactive part depends on the dipole thickness. This is equal to the slot width, designated w. Even with the sinusoidal approximation of $I_1(x_2)$, the expression for $Z(x_1)$ is rather complicated. It uses the sine and modified cosine integral functions Si(x) and Cin(x), defined as follows:

$$\text{Si}(x) = \int_0^x \frac{\sin u}{u} \, du \tag{10-18}$$

$$\text{Cin}(x) = \int_0^x \frac{1 - \cos u}{u} \, du \tag{10-19}$$

---

2. Edward C. Jordan and Keith G. Balmain, *Electromagnetic Waves and Radiating Systems,* 2nd ed. (Englewood Cliffs, N.J.: Prentice Hall Press, 1968), p. 326. Adapted by permission of Prentice Hall Press, Englewood Cliffs, New Jersey.

Despite the apparent simplicity of these integrals, they are not expressible in terms of elementary functions. Tables are available, or they may be computed using Simpson's rule or other numerical methods.

Each section of the dipole behaves as a monopole antenna, and unless $x_1 = 0$, the two sections are of different lengths. Each monopole has a driving-point impedance that depends on its length and the thickness w. The total dipole impedance $Z(x_1)$ is then equal to the two monopole impedances in series. For a monopole of length h, the real and imaginary parts of its driving-point impedance $Z_m(h, w)$ are given by[3]

$$R_m(h) = \frac{15\{[2 + 2\cos(2\beta h)] \, \text{Cin}(2\beta h) - \cos(2\beta h) \, \text{Cin}(4\beta h) - 2\sin(2\beta h) \, \text{Si}(2\beta h) + \sin(2\beta h) \, \text{Si}(4\beta h)\}}{1 - \cos(2\beta h)} \quad \textbf{(10-20)}$$

$$X_m(h, w) = 15 \left\{ \sin(2\beta h) \left[-4 \sinh^{-1}\left(\frac{2h}{w}\right) + 2 \sinh^{-1}\left(\frac{4h}{w}\right) + \right.\right.$$
$$2 \, \text{Cin}(v_1) - 2 \, \text{Cin}(u_1) + \text{Cin}(v_2) - \text{Cin}(u_2)] +$$
$$\cos(2\beta h)[2 \, \text{Si}(u_1) - 2 \, \text{Si}(\beta w/2) + 2 \, \text{Si}(v_1) -$$
$$\text{Si}(u_2) - \text{Si}(v_2)] + 2[\text{Si}(u_1) - 2 \, \text{Si}(\beta w/2) + \text{Si}(v_1)]\} \quad \textbf{(10-21)}$$

where

$$u_1 = \beta[\sqrt{h^2 + (w/2)^2} - h] \qquad u_2 = \beta[\sqrt{(2h)^2 + (w/2)^2} + 2h]$$
$$v_1 = \beta[\sqrt{h^2 + (w/2)^2} + h] \qquad v_2 = \beta[\sqrt{(2h)^2 + (w/2)^2} - 2h]$$

Graphs of these functions, for various dipole thicknesses, appear in Figure 10-9, and this accuracy is adequate for our use. The exact formulas are given for use in a computer if desired.

The two monopole antennas are effectively connected in series with the source V_1. The driving-point impedance $Z_1(x_1)$ of the dipole antenna at point x_1 is therefore the sum of the impedances of the two monopoles, or

$$Z_1(x_1) = Z_m(l/2 + x_1) + Z_m(l/2 - x_1) \quad \textbf{(10-22)}$$

Knowing $Z_1(x_1)$ gives us the transfer admittance $y(x_2, x_1)$, which is

$$y(x_2, x_1) = \frac{I_1(x_2)}{V_c} = \frac{I_1(x_2)}{I_1(x_1)Z_1(x_1)}$$

$$= \begin{cases} \dfrac{I_1(x_1) \sin\left[\beta\left(\dfrac{l}{2} + x_2\right)\right]}{\left[Z_m\left(\dfrac{l}{2} + x_1\right) + Z_m\left(\dfrac{l}{2} - x_1\right)\right] \sin\left[\beta\left(\dfrac{l}{2} + x_1\right)\right]} & \text{for } -\dfrac{l}{2} \le x_2 \le x_1 \\[30pt] \dfrac{I_1(x_1) \sin\left[\beta\left(\dfrac{l}{2} - x_2\right)\right]}{\left[Z_m\left(\dfrac{l}{2} + x_1\right) + Z_m\left(\dfrac{l}{2} - x_1\right)\right] \sin\left[\beta\left(\dfrac{l}{2} - x_1\right)\right]} & \text{for } x_1 \le x_2 \le \dfrac{l}{2} \end{cases} \quad \textbf{(10-23)}$$

3. Edward C. Jordan and Keith G. Balmain, *Electromagnetic Waves and Radiating Systems*, 2nd ed. (Englewood Cliffs, N.J.: Prentice Hall Press, 1968), pp. 540–47. Adapted by permission of Prentice Hall Press, Englewood Cliffs, New Jersey.

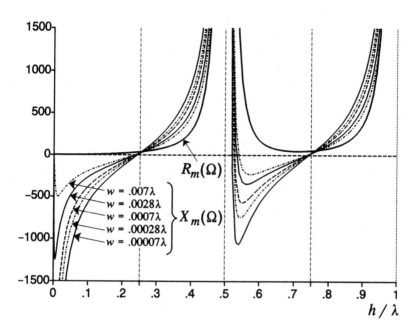

Figure 10-9 Driving-point impedance vs. monopole antenna length

Now we activate all sources in Figure 10-8, and the current distribution changes. Each source contributes to the current $I(x_1)$ at point x_1. The contribution from the source at point x_2 is $Vy(x_1, x_2)$, which by reciprocity is equal to $Vy(x_2, x_1)$. If the sources are due to the approaching field $\mathbf{E}_0$, and if dx is the distance between sources, then the voltage of each source is $dV = |\mathbf{E}_0|_s \, dx$, where $|\mathbf{E}_0|_s$ is the spatial magnitude of $(\mathbf{E}_0)_s$. The total current $I(x_1)$ at point x_1 is then

$$I(x_1) = \int_l y(x_1, x_2) \, dV = |\mathbf{E}_0|_s \int_{-l/2}^{l/2} y(x_1, x_2) \, dx_2$$

$$= \frac{|\mathbf{E}_0|_s}{Z_m(l/2 + x_1) + Z_m(l/2 - x_1)} \cdot$$

$$\left\{ \frac{\displaystyle\int_{-l/2}^{x_1} \sin[\beta(l/2 + x_2)] \, dx_2}{\sin[\beta(l/2 + x_1)]} + \frac{\displaystyle\int_{x_1}^{-l/2} \sin[\beta(l/2 - x_2)] \, dx_2}{\sin[\beta(l/2 - x_1)]} \right\}$$

$$= \frac{|\mathbf{E}_0|_s}{Z_m(l/2 + x_1) + Z_m(l/2 - x_1)} \cdot \left\{ \frac{1 - \cos[\beta(l/2 + x_1)]}{\beta \sin[\beta(l/2 + x_1)]} + \frac{1 - \cos[\beta(l/2 - x_1)]}{\beta \sin[\beta(l/2 - x_1)]} \right\}$$

$$= \frac{|\mathbf{E}_0|_s \{ \tan[\beta(l/2 + x_1)/2] + \tan[\beta(l/2 - x_1)/2] \}}{\beta[Z_m(l/2 + x_1) + Z_m(l/2 - x_1)]}$$

$$= \frac{|\mathbf{E}_0|_s \lambda \{ \tan[\pi(l/2 + x_1)/\lambda] + \tan[\pi(l/2 - x_1)/\lambda] \}}{2\pi[Z_m(l/2 + x_1) + Z_m(l/2 - x_1)]} \qquad \textbf{(10-24)}$$

At some points on the dipole, the tangent function in Equation (10-24) becomes infinite, but so does $Z_m(l/2 + x_1) + Z_m(l/2 - x_1)$. At those points Equation (10-24) becomes indeterminate and is not usable. This occurs at only a finite number of points, so it will cause no difficulty.

The current on the dipole antenna is equivalent to many small oscillating dipoles. For each dipole of length dx, the relation between the current $I(x_1)$ and the dipole moment $Q\,dx$ is

$$I(x_1)dx = \frac{\partial}{\partial t}Qdx = j\omega Qdx \tag{10-25}$$

From Equation (8-37), the reradiated far field $d(\mathbf{E}_t)_s$ due to an infinitesimal current element $I(x_1)dx$ at the point x_1 is

$$d(\mathbf{E}_t)_s = -\frac{\beta^2 Qdxe^{-j\beta r}}{4\pi\epsilon_0 r}\sin\theta\,\mathbf{a}_\theta = -\frac{I(x_1)dx}{j\omega}\cdot\frac{\beta^2 e^{-j\beta r}}{4\pi\epsilon_0 r}\sin\theta\,\mathbf{a}_\theta$$

$$= -\frac{I(x_1)dx}{j\omega}(\omega\sqrt{\mu_0\epsilon_0})\left(\frac{2\pi}{\lambda}\right)\frac{e^{-j\beta r}}{4\pi\epsilon_0 r}\sin\theta\,\mathbf{a}_\theta$$

$$= \frac{j\eta_0 I(x_1)dxe^{-j\beta r}}{2\lambda r}\sin\theta\,\mathbf{a}_\theta \tag{10-26}$$

The direction of maximum radiation is broadside to the dipole antenna, or $\theta = \pi/2$. In this direction the time magnitude of the reradiated field is

$$d|(\mathbf{E}_t)_s| = \frac{\eta_0|I(x_1)|dx}{2\lambda r}\mathbf{a}_\theta$$

$$= \frac{\eta_0}{2\lambda r}\left\{\frac{\|\mathbf{E}_0\|\lambda\{\tan[\pi(l/2 + x_1)/\lambda] + \tan[\pi(l/2 - x_1)/\lambda]\}}{2\pi|Z_m(l/2 + x_1) + Z_m(l/2 - x_1)|}\right\}dx\,\mathbf{a}_\theta$$

$$= \frac{\|\mathbf{E}_0\|\eta_0\{\tan[\pi(l/2 + x_1)/\lambda] + \tan[\pi(l/2 - x_1)/\lambda]\}}{4\pi r|Z_m(l/2 + x_1) + Z_m(l/2 - x_1)|}dx\,\mathbf{a}_\theta \tag{10-27}$$

Also, in this direction, all current elements are approximately equidistant from the measurement point, so the generated fields add in phase. The total field is the integral of these infinitesimal elements:

$$|(\mathbf{E}_t)_s| = \int_l d|(\mathbf{E}_t)_s|$$

$$= \int_{-l/2}^{l/2}\frac{\|\mathbf{E}_0\|\eta_0\{\tan[\pi(l/2 + x_1)/\lambda] + \tan[\pi(l/2 - x_1)/\lambda]\}}{4\pi r|Z_m(l/2 + x_1) + Z_m(l/2 - x_1)|}dx_1\,\mathbf{a}_\theta \tag{10-28}$$

where $Z_m(l/2 + x_1)$ and $Z_m(l/2 - x_1)$ are given by Equations (10-20) and (10-21).

The sine and modified cosine integral functions in these expressions complicate the evaluation of this integral. We could evaluate it by numerical integration with a sufficiently fast computer. For our purposes, it will be sufficient to estimate it by inspection of the current distribution $I(x_1)$. This distribution is shown in

Figure 10-10 for $l \leq \lambda/2$ and in Figure 10-11 for $l > \lambda/2$, where in both figures $w = 0.0028\lambda$. For $l = \lambda$, $I(x_1)$ is identically zero and therefore its distribution is not visible in the figure. The distributions resemble one or more half cycles of a sine wave, particularly those for $l < \lambda/2$, which are the cases of greatest interest. For a true sine wave, the average value is $2/\pi$ times the maximum, and we will assume this ratio for the current distributions in Figures 10-10 and 10-11. The maximum value, or close to it, is at the center of the dipole, or $x_1 = 0$. The integral is equal to the average of $I(x_1)$ multiplied by l. With our simplifying approximation, this is $2l/\pi$ times the current at the center, or

$$\int_{-l/2}^{l/2} I(x_1)\,dx_1 \approx \frac{2I(0)l}{\pi} \tag{10-29}$$

With this approximation, Equation (10-28) becomes

$$|(\mathbf{E}_t)_s| = \int_l d|(\mathbf{E}_t)_s| = \int_{-l/2}^{l/2} \frac{\eta_0 |I(x_1)|\,dx}{2\lambda r}\mathbf{a}_\theta \approx \frac{2l}{\pi}\cdot\frac{\eta_0 |I(0)|}{2\lambda r}\mathbf{a}_\theta$$

$$= \frac{l\eta_0}{\pi\lambda r}\left\{\frac{\|\mathbf{E}_0\|\lambda\left[\tan\left(\dfrac{\pi l}{2\lambda}\right)+\tan\left(\dfrac{\pi l}{2\lambda}\right)\right]}{2\pi|Z_m(l/2)+Z_m(l/2)|}\right\}\mathbf{a}_\theta$$

$$= \frac{\|\mathbf{E}_0\|l\eta_0}{2\pi^2 r|Z_m(l/2)|}\tan\left(\frac{\pi l}{2\lambda}\right)\mathbf{a}_\theta \tag{10-30}$$

For a slot instead of a dipole, because of the duality mentioned above, an identical equation results, except that $\eta_0\mathbf{H}_0$ replaces $\mathbf{E}_0$, and $\eta_0\mathbf{H}_t$ replaces $\mathbf{E}_t$. The graphs in Figures 10-10 and 10-11 then represent the voltage across the slot at various points. The equation corresponding to Equation (10-30) is

$$|(\mathbf{H}_t)_s| = \frac{\|\mathbf{H}_0\|l\eta_0}{2\pi^2 r|Z_m(l/2)|}\tan\left(\frac{\pi l}{2\lambda}\right)\mathbf{a}_\theta \tag{10-31}$$

where $\eta_0 = 120\pi$ ohms and $Z_m(l/2)$ is as given by Equations (10-20) and (10-21) for a dipole of the same dimensions as the slot. A graph of this equation, for $w = 0.0028\lambda$, appears in Figure 10-12. For a 150-MHz signal, for example, this value of w equates to a slot width of 5.6 mm. For narrower slots, Figure 10-9 shows that the reactive part of Z_m increases, except for l equal to odd multiples of $\lambda/2$. The voltage across the slot, and the reradiated field, thus decrease, but Figure 10-12 still provides an upper bound. Due to the complexity of Equations (10-20) and (10-21), and the unfamiliar functions, it is usually preferable to read the attenuation from the graph. The analytical expression, however, is useful in computer programs.

From Figure 10-12 we note that for a slot of length $\lambda/2$, the reradiated signal strength is nearly 30 percent of the incident signal strength, at a distance $r = \lambda$. For this slot length, Z_m is resistive and therefore w has little effect. Thus, a half-wave slot greatly degrades the shielding effect. Although some longer slot lengths provide greater attenuation at certain frequencies, this effect is normally useless because

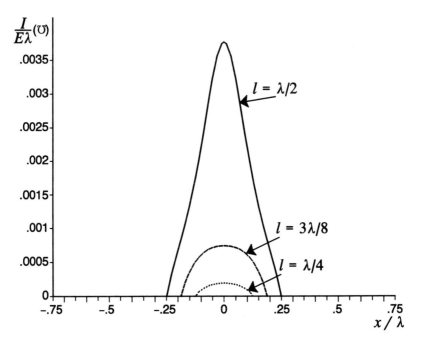

Figure 10-10   Current distribution on a receiving antenna, $l \le \lambda/2$

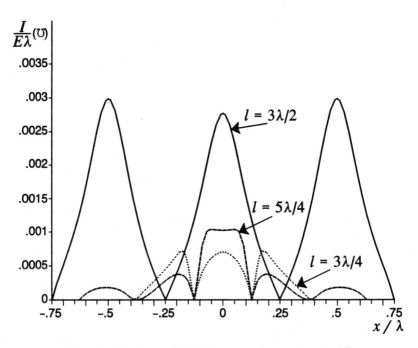

Figure 10-11   Current distribution on a receiving antenna, $l > \lambda/2$

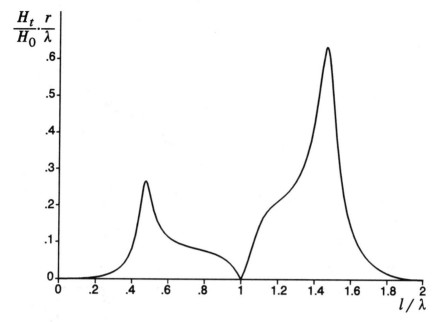

Figure 10-12 Reradiated field versus length of slot in shield

several frequencies are usually present. For a shield to be effective, any slots must be shorter than $\lambda/2$ at the highest frequency present. A common practice is to assume that the signal reradiated from a slot shorter than $\lambda/2$ is proportional to the slot length, with zero attenuation for a half-wave slot. By examining the curve in Figure 10-12 we can derive a more useful approximation.

If we plot this portion of the curve on a semi-log grid, the graph appears as in Figure 10-13. A large interval on the graph is nearly linear, which means that the function is nearly exponential over that interval. Let

$$\hat{H}_t = \|\mathbf{H}_0\| \frac{\lambda}{r} \cdot \frac{e^{15.25l/\lambda}}{4500} \tag{10-32}$$

Then the graph of $\hat{H}_t$ on a semi-log grid is the straight line shown in Figure 10-13. From this approximation, the shielding effectiveness for the slot of width equal to 0.0028λ is

$$S = 20 \log \frac{\|\mathbf{H}_0\|}{\hat{H}_t}$$

$$= 20 \log \frac{r}{\lambda} - 15.25 \frac{l}{\lambda} \cdot 20 \log e + 20 \log 4500$$

$$= 73 - 132.45 \frac{l}{\lambda} + 20 \log \frac{r}{\lambda} \ \text{dB} \tag{10-33}$$

With this approximation, the shielding effectiveness of a half-wave slot ($l = \lambda/2$) at a distance $r = \lambda$ is 6.78 dB. This is a bit more realistic than the 0 dB sometimes

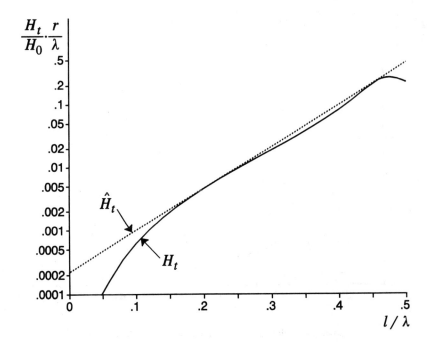

Figure 10-13   Approximation of reradiated field from shield slot

assumed. For $l < 0.1\lambda$, $\hat{H}_t$ is too high, as is evident from Figure 10-13. Obviously $\hat{H}_t$ does not become zero for $l = 0$, as it should. The approximation is adequate for our purposes, however, and we will use it from now on. Where it does differ from the true function, the error is on the high or "safe" side. Equation (10-33) would be different for other slot widths. For narrower slots, the shielding effectiveness improves and is never less than that calculated above.

The preceding analysis considers only a single slot. With multiple slots, the combined effect depends on the relative phases of the fields emanating from the various slots. It is possible that they all may be in phase. The resultant field is then the vector sum of the time magnitudes of the fields caused by the individual slots. At certain angles, the radiated fields may even be stronger than with no shield at all. The slots behave as a multi-element array, which is obviously very undesirable. The directive pattern narrows the region of probable interference, but a sensitive device may happen to be located in a lobe of strong radiation. We will note in Chapter 12 that regulations which limit permissible electromagnetic radiation relate to the direction of maximum field strength. If an array of slots causes a highly directional radiation pattern, the maximum field strength will be higher than if the radiation were uniform in all directions. This increases the likelihood that the equipment will violate the regulations and should be avoided.

We will now calculate the field radiated from multiple slots. We assume that the slots are far enough apart so that no slot interferes with the aperture current $\mathbf{J}_a$ from an adjacent slot. We will first consider the case of two slots, and then generalize it to an arbitrary number. The sum E_t of two fields E_1 and E_2 of equal magnitude E but differing by a phase angle θ is

$$E_t = E_1 + E_2 = E + Ee^{j\theta} = E(1 + e^{j\theta})$$

$$|E_t| = E\sqrt{(1 + \cos\theta)^2 + \sin^2\theta}$$

$$= E\sqrt{1 + 2\cos\theta + 1} = E\sqrt{2 + 2\left(2\cos^2\frac{\theta}{2} - 1\right)}$$

$$= 2E\left|\cos\frac{\theta}{2}\right| \tag{10-34}$$

The maximum field strength due to these two slots occurs when the fields are in phase, or when $\theta = 0$ and $\cos(\theta/2) = 1$. Then the resulting field strength is $2E$. An extension of this reasoning would show that, for n identical slots instead of two, with all phases equal, the field strength is nE. Thus, to account for multiple slots with all fields at the same phase angle, we add the following term to Equation (10-33):

$$S_m = -20\log\frac{nE}{E} = -20\log n \tag{10-35}$$

This term is in addition to all other effects. It is the worst case, since if not all fields are in phase, the resulting field will be weaker and the attenuation will be greater.

More usually the relative phases at the various slots will be random. To analyze the effects of a random phase difference, we assume the phase θ to be a random variable with uniform distribution from 0 to 2π. The calculated total field strength will then be an estimate. It will be a poor estimate for two slots, but it improves as the number of slots increases. Depending on the equipment affected by the fields, either the absolute mean or the root-mean-square of the individual field strengths may be a more meaningful estimate. The absolute mean is

$$\overline{|E_t|} = \frac{1}{2\pi}\int_0^{2\pi}|E_t|d\theta$$

$$= \frac{1}{2\pi}\int_0^{2\pi}2E\left|\cos\frac{\theta}{2}\right|d\theta = \frac{1}{\pi}\int_0^{\pi}2E|\cos\phi|d\phi$$

$$= \frac{4E}{\pi}\int_0^{\pi/2}\cos\phi\,d\phi = \frac{4E}{\pi}\sin\phi\Big|_0^{\pi/2} = \frac{4E}{\pi} \tag{10-36}$$

The root-mean-square is

$$\sqrt{\overline{|E_t|^2}} = \sqrt{\frac{1}{2\pi}\int_0^{2\pi}|E_t|^2\,d\theta}$$

$$= \sqrt{\frac{1}{2\pi}\int_0^{2\pi}2E^2(1 + \cos\theta)\,d\theta}$$

$$= \sqrt{\frac{2E^2}{2\pi}(\theta + \sin\theta)\Big|_0^{2\pi}} = \sqrt{2}E \tag{10-37}$$

The root-mean-square gives a more conservative figure and is therefore the figure normally used. For the more general case of n slots, Equation (10-37) is generalized:

$$\sqrt{\overline{|E_t|^2}} = \sqrt{n}\, E \tag{10-38}$$

Thus, to account for multiple slots with the fields at random phases, we add the following term to Equation (10-33):

$$S_m = -20 \log \frac{\sqrt{\overline{|E_t|^2}}}{E} = -20 \log \sqrt{n} = -10 \log n \tag{10-39}$$

Like Equation (10-35), this term is in addition to all other effects. This expression is an estimate, assuming random, independent phases, and assuming that the slots are sufficiently separated so that their aperture currents do not interact.

10-3 CRITICAL DIMENSIONS OF APERTURES

Throughout this analysis we have assumed that the E-field polarization is perpendicular to the slot. For this polarization, the width of the slot is much less important than the length. This is consistent with our earlier observation comparing Figures 10-2 and 10-3, where the slot evidently degrades the shielding effectiveness nearly as much as the round hole.

If, instead, the E field is parallel to the slot, the shorter dimension becomes the width in the above calculations, and the leakage through the slot decreases. Since we usually cannot predict the wave polarization, we normally assume the worst case. This occurs with the longest dimension of the slot perpendicular to the current, as assumed above. Therefore, the longest dimension of an aperture is the most critical and is the dimension that we must minimize.

Comparing Equations (10-33) and (10-39) shows the improvement obtained by using several small slots instead of one large one. Two identical slots increase the expected leakage by 3 dB over that of either slot if the relative phase is random, or 6 dB if the fields are in phase. Doubling the length l of a single slot, if it remains shorter than $\lambda/2$, increases the leakage by $132l/\lambda$ dB, which is much worse unless $l < \lambda/20$. The importance of keeping slots short is obvious.

It is easy to overlook the effect of a slot in a shield. At a joint between two panels of a shield, slots appear between adjacent screws or rivets, unless the joint is continuously welded or soldered. Overlapping the panels without electrically bonding them does not eliminate the effects of the slot, for then the potential shown in Figure 10-6 still exists across the slot. The capacitance between the overlapped panels will allow some displacement current to flow across the slot, so the emissions may be less than they would be with no overlap. However, this displacement current is effectively an oscillating electric dipole, which then acts as a new field source. For the analysis of Section 10-2 to be valid, a *conduction* current **J** must flow around the ends of the slot. This requires metallic bonding between the panels. The bonding points between the panels should be spaced as closely as possible, since this spacing is the distance *l* used in the calculations of Section 10-2. It must be short compared to a $\lambda/2$ at the highest frequency generated by the equipment.

A particularly troublesome case is a panel that must be opened for servicing. For convenience, such panels are often secured by one screw at each corner. Then each edge may act as a slot antenna, if the equipment uses high enough frequencies. The slots may be shortened by making the mating surfaces conductive and ensuring that they touch. One approach is to use a plating of tin, nickel, or cadmium, instead of paint. Sometimes conductive gaskets may be necessary. They must have a very low resistivity; some conductive gaskets designed solely for electrostatic protection are not suitable for EMI shield bonding.

If a metal shield cannot be used, perhaps for economic or aesthetic reasons, another conductive material may be feasible. Conductive paint can be used on plastic cabinets and has been used to shield entire rooms. It may be applied like any paint, but requires constant mixing since it contains as much as 80 percent powdered nickel. As with any shield, it is important to bond all joints; otherwise, they act as slot antennas. A process known as flame spraying or arc spraying provides a more durable conductive finish and can be applied to cabinets or room walls. The process involves spraying molten metal on the surfaces, which cools rapidly enough that it causes no damage. Decorative paint may be applied over either type of metallization.

For an application such as a cathode ray tube (*CRT*) display, a wire screen can be mounted over the face of the tube. However, a wire screen is unattractive, reduces the display brightness, and causes optical interference ("moiré") patterns on a color CRT display. To prevent leakage from slots around the wire screen, it must be well bonded to the cabinet. An alternative is to provide a funnel-shaped shield behind the CRT, which prevents the interfering fields from reaching the CRT face. Then the only possible source of interference is the CRT itself. A CRT uses high voltages and low currents, so it is usually an electric field source instead of magnetic. The face of the CRT is in the near electric field, where there is very little magnetic field strength. To reduce the electric field strength, we may apply a transparent conductive coating to the face of the CRT, or we may mount a conductive glass or plastic shield over it. These also reduce the display brightness, but not as severely as a wire screen. The conductivity of a clear coating or shield is usually insufficient to protect against magnetic fields, but it is adequate for an electric field source such as a CRT. The clear coating or shield must be grounded to the cabinet but, to block electric fields, only one ground connection is necessary, as discussed in Chapter 4. Additional bonding points, of course, do help somewhat.

10-4 WAVEGUIDE THEORY

When an aperture is unavoidable, it may be possible to lessen its effect by mounting a waveguide in it, as shown in Figure 10-14. A waveguide can propagate electro-

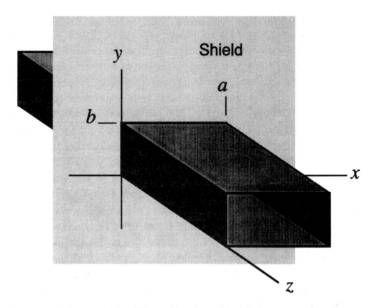

Figure 10-14 Waveguide mounted in a shield aperture

magnetic energy only above a certain frequency that depends on its cross-sectional dimensions. This frequency is known as the *cutoff* frequency. Below that frequency, an electromagnetic field passing through the waveguide decreases very rapidly with distance. If the waveguide is small enough that the cutoff frequency is greater than all significant frequencies generated by the equipment, then all radiated emissions will be attenuated. The amount of attenuation depends on the length of the waveguide.

We first consider a rectangular waveguide of width a and height b, oriented in a rectangular coordinate system as shown in Figure 10-14. The **E** and **H** fields inside the waveguide must, of course, satisfy Maxwell's equations. At the boundaries $x = 0$, $x = a$, $y = 0$, and $y = b$, the tangential components of **E** and the normal components of **H** must be zero. The solutions belong to either of two forms, known as *TE* (*T*ransverse *E*lectric) and *TM* (*T*ransverse *M*agnetic). The *TE* solutions are of the form

$$(\mathbf{E}_{mn})_s = \frac{j\omega\mu_0 K e^{-\gamma_{mn}z}}{A_n^2 + B_m^2}[A_n \cos(B_m x) \sin(A_n y)\,\mathbf{i} - B_m \sin(B_m x) \cos(A_n y)\,\mathbf{j} \quad \textbf{(10-40)}$$

$$(\mathbf{H}_{mn})_s = K e^{-\gamma_{mn}z}\left\{\frac{\gamma_{mn}}{A_n^2 + B_m^2}[B_m \sin(B_m x) \cos(A_n y)\,\mathbf{i} + \right.$$
$$\left. A_n \cos(B_m x) \sin(A_n y)\,\mathbf{j}] + \cos(B_m x) \cos(A_n y)\,\mathbf{k}\right\} \qquad \textbf{(10-41)}$$

where m and n are arbitrary integers, and

$$B_m = \frac{m\pi}{a} \qquad A_n = \frac{n\pi}{b} \tag{10-42}$$

$$\gamma_{mn} = \sqrt{B_m^2 + A_n^2 - \omega^2 \mu_0 \epsilon_0} = \sqrt{\left(\frac{m\pi}{a}\right)^2 + \left(\frac{n\pi}{b}\right)^2 - \omega^2 \mu_0 \epsilon_0} \tag{10-43}$$

The *TM* solutions are of the form

$$(\mathbf{E}_{mn})_s = -Ke^{-\gamma_{mn}z}\left\{\frac{\gamma_{mn}}{A_n^2 + B_m^2}[B_m \cos(B_m x)\sin(A_n y)\,\mathbf{i} + \right.$$

$$\left. A_n \sin(B_m x)\cos(A_n y)\,\mathbf{j}] + \sin(B_m x)\sin(A_n y)\,\mathbf{k}\right\} \tag{10-44}$$

$$(\mathbf{H}_{mn})_s = \frac{j\omega\epsilon_0 Ke^{-\gamma_{mn}z}}{A_n^2 + B_m^2}[A_n \sin(B_m x)\cos(A_n y)\,\mathbf{i} - B_m \cos(B_m x)\sin(A_n y)\,\mathbf{j} \tag{10-45}$$

where A_n, B_m, and γ_{mn} are the same as for the *TE* solutions.

The values of m and n determine the *mode* of the wave in the waveguide. They are written as subscripts to identify the modes, which then appear as TE_{11}, TM_{12}, and so on. Generally, there may be several different modes present at once, in which case the fields are the sums of the expressions for the different modes.

As the measurement point moves along the waveguide (in the z direction), the field intensities vary depending on the $e^{-\gamma z}$ factor in the field expressions. From Equation (10-43) we note that γ_{mn} is either purely real or purely imaginary, depending on the sign of the radicand. If γ_{mn} is imaginary, the phases of $\mathbf{E}$ and $\mathbf{H}$ are retarded in proportion to z but their amplitudes remain constant. If γ_{mn} is real, their amplitudes decrease exponentially with z.

For a given mode, the sign of the radicand in Equation (10-43) depends on the frequency. The cutoff frequency f_c is the frequency for which the radicand is zero. This frequency is

$$\left(\frac{m\pi}{a}\right)^2 + \left(\frac{n\pi}{b}\right)^2 - (2\pi f_c)^2 \mu_0 \epsilon_0 = 0$$

$$f_c = \sqrt{\frac{(m\pi/a)^2 + (n\pi/b)^2}{(2\pi)^2 \mu_0 \epsilon_0}} = \sqrt{\frac{(m/a)^2 + (n/b)^2}{4\mu_0 \epsilon_0}} \tag{10-46}$$

Above this frequency, γ_{mn} is imaginary; below it, γ_{mn} is real. Therefore, any frequency below f_c will be attenuated exponentially in the z direction.

The cutoff frequency depends on the mode, and it is difficult to control the mode or modes present in the waveguide. Therefore, we can only be sure that the field will be attenuated if its frequency is below the lowest cutoff frequency of all modes. From Equation (10-46) it is obvious that the lowest cutoff frequency will be for the lowest possible values of m and n, but we must investigate what values are possible.

If both m and n are zero, all fields become zero in both TE and TM modes, so this combination is not possible. If either m or n is zero in the TM mode, the fields also are zero. They can be nonzero in the TE mode, however, if either m or n (but not both) is zero. Therefore, the lowest cutoff frequency occurs for the TE_{10} or the TE_{01} mode. By convention, a rectangular waveguide is normally oriented so that its longer transverse dimension is a (that is, in the x direction). Then the lowest cutoff frequency occurs for the TE_{10} mode, known as the *dominant* mode. This frequency is

$$f_c = \sqrt{\frac{(1/a)^2 + (0/b)^2}{4\mu_0\epsilon_0}} = \frac{1}{2a}\frac{1}{\sqrt{\mu_0\epsilon_0}} = \frac{c}{2a}$$

$$= \frac{1.5 \cdot 10^4 \text{ cm}/\mu\text{sec}}{a} \tag{10-47}$$

Below this frequency, the attenuation in the z direction through a waveguide of length l is

$$\frac{|\mathbf{E}_0|_s}{|\mathbf{E}_t|_s} = e^{\gamma l} = \exp\left[l\sqrt{\left(\frac{\pi}{a}\right)^2 - \omega^2\mu_0\epsilon_0}\right]$$

$$= \exp\left[\pi\sqrt{\left(\frac{l}{a}\right)^2 - \left(\frac{2f}{c}\right)^2}\right] \tag{10-48}$$

If $f \ll f_c$, then $2f/c$ becomes insignificant in Equation (10-48), which then reduces to

$$\frac{|\mathbf{E}_0|_s}{|\mathbf{E}_t|_s} = e^{\pi l/a} \tag{10-49}$$

The attenuation expressed in decibels, for $f \ll f_c$, is

$$S_w = 20 \log \frac{|\mathbf{E}_0|_s}{|\mathbf{E}_t|_s} = 20 \log e^{\pi l/a}$$

$$= (20\pi \log e)\frac{l}{a} = 27.3\frac{l}{a} \text{ dB} \tag{10-50}$$

We next consider a cylindrical waveguide of radius a, centered on the z axis of a cylindrical coordinate system. This analysis is complicated by nonelementary functions. The solutions to Maxwell's equations with cylindrical boundary conditions involve Bessel functions, which are not expressible in terms of sines and cosines. They are tabulated in most mathematical tables, and their derivatives are expressible in terms of other Bessel functions. A Bessel function of the kind encountered in a cylindrical waveguide is designated $J_n(x)$, where n is the *order* of the function. Graphs of Bessel functions of order 0 through 4 appear in Figure 10-15.

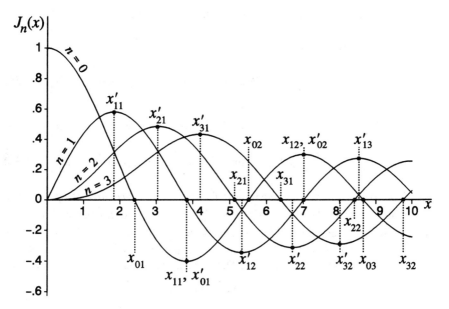

Figure 10-15   Bessel functions and their roots

To satisfy the boundary conditions we must scale the function arguments such that the value of the function is zero at the boundary. To do so, we must know the values of x that cause $J_n(x)$ to be zero, that is, the roots of $J_n(x)$. These are transcendental numbers and must be obtained from tables, just like the functions themselves. The mth root of $J_n(x)$ is designated x_{nm}. The roots of the functions in Figure 10-15 also appear in the figure.

The **E** and **H** fields inside the waveguide must satisfy Maxwell's equations and the boundary conditions at $\rho = a$. As is true for the rectangular waveguide, the solutions are either *TE* or *TM*. The *TM* solutions are of the form

$$(\mathbf{E}_{nm})_s = Ke^{-\gamma_{nm}z}\left\{\frac{\gamma_{nm}}{A_{nm}}\left[-J'(A_{nm}\rho)\cos(n\phi)\,\mathbf{a}_\rho + \right.\right.$$

$$\left.\left.\frac{n}{A_{nm}\rho}J_n(A_{nm}\,\rho)\sin(n\phi)\,\mathbf{a}_\phi\right] + J_n(A_{nm}\rho)\cos(n\phi)\,\mathbf{k}\right\} \quad \textbf{(10-51)}$$

$$(\mathbf{H}_{nm})_s = -\frac{j\omega\epsilon_0 Ke^{-\gamma_{nm}z}}{A_{nm}}\left[\frac{n}{A_{nm}\rho}J_n(A_{nm}\rho)\sin(n\phi)\,\mathbf{a}_\rho + J'_n(A_{nm}\rho)\cos(n\phi)\,\mathbf{a}_\phi\right] \quad \textbf{(10-52)}$$

where m and n are arbitrary integers, $J'_n(A_{nm}\rho)$ is the derivative of $J_n(A_{nm}\rho)$ and

$$A_{nm} = \frac{x_{nm}}{a} = \frac{\text{the } m\text{th root of } J_n(x)}{a} \quad \textbf{(10-53)}$$

$$\gamma_{nm} = \sqrt{A_{nm}^2 - \omega^2\mu_0\epsilon_0} \quad \textbf{(10-54)}$$

The roots of $J_n(x)$ cause $J_n(A_{nm}a)$ to be zero, and therefore E_ϕ and H_ρ are zero for $\rho = a$, as required.

The *TE* solutions are of the form

$$(\mathbf{E}_{nm})_2 = -\frac{j\omega\mu_0 K e^{-\gamma_{mn}z}}{B_{nm}}\left[\frac{n}{B_{nm}\rho}J_n(B_{nm}\rho)\sin(n\phi)\,\mathbf{a}_\rho + J_n'(B_{nm}\rho)\cos(n\phi)\,\mathbf{a}_\phi\right] \quad \textbf{(10-55)}$$

$$(\mathbf{H}_{nm})_s = K e^{-\gamma_{mn}z}\left\{\frac{\gamma_{mn}}{B_{nm}}\left[-J'(B_{nm}\rho)\cos(n\phi)\,\mathbf{a}_\rho +\right.\right.$$

$$\left.\left.\frac{n}{B_{nm}\rho}J_n(B_{nm}\rho)\sin(n\phi)\,\mathbf{a}_\phi\right] + J_n(B_{nm}\rho)\cos(n\phi)\,\mathbf{k}\right\} \quad \textbf{(10-56)}$$

Here $J_n'(B_{nm}a)$ must be zero, so we require the roots of the *derivative* of $J_n(x)$ instead of the function itself. These are designated x_{nm}' and are also transcendental numbers that must be obtained from tables. They also appear in Figure 10-15 and are the points of zero slope. One of these values will appear in the formula for the cutoff frequency of a cylindrical waveguide. The field expressions for the *TE* modes are

$$B_{nm} = \frac{x_{nm}'}{a} = \frac{\text{the } m\text{th root of } J_n'(x)}{a} \quad \textbf{(10-57)}$$

$$\gamma_{nm} = \sqrt{B_{nm}^2 - \omega^2\mu_0\epsilon_0} \quad \textbf{(10-58)}$$

As with the rectangular waveguide, the cutoff frequency is the frequency for which the radicand of Equation (10-54) or (10-58) is zero. The mode having the lowest cutoff frequency is the mode with the smallest value of A_{nm} or B_{nm}. Inspection of Figure 10-15 shows that the smallest root of $J_n'(x)$ is x_{11}', which is smaller than any root of $J_n(x)$. Its tabulated value is 1.841. The dominant mode of a cylindrical waveguide is therefore the TE_{11} mode, and its cutoff frequency is

$$h_{11}'^2 - (2\pi f_c)^2\mu_0\epsilon_0 = 0$$

$$f_c = \sqrt{\frac{h_{11}'^2}{(2\pi)^2\mu_0\epsilon_0}} = \frac{h_{11}'}{2\pi\sqrt{\mu_0\epsilon_0}} = \frac{x_{11}'c}{2\pi a}$$

$$= \frac{8.79}{a}\text{ GHz} \quad \textbf{(10-59)}$$

with a in centimeters.

Below this frequency, the attenuation in the z direction through a cylindrical waveguide of length l is

$$\frac{|\mathbf{E}_0|_s}{|\mathbf{E}_t|_s} = e^{\gamma l} = \exp\left[l\sqrt{h_{11}'^2 - \omega^2\mu_0\epsilon_0}\right]$$

$$= \exp\left[\sqrt{\left(\frac{x_{11}l}{a}\right)^2 - \left(\frac{2\pi f}{c}\right)^2}\right] \quad \textbf{(10-60)}$$

If $f \ll f_c$, then $2\pi f/c$ becomes insignificant in Equation (10-60), which then reduces to

$$\frac{|\mathbf{E}_0|_s}{|\mathbf{E}_t|_s} = e^{x_{11}l/a} \qquad (10\text{-}61)$$

The attenuation expressed in decibels, for $f \ll f_c$, is

$$S_w = 20 \log \frac{|\mathbf{E}_0|}{|\mathbf{E}_t|} = 20 \log e^{x_{11}l/a} = (20x_{11} \log e)\frac{l}{a}$$

$$= 16.0 \frac{l}{a} \text{ dB} = 32.0 \frac{l}{d} \text{ dB} \qquad (10\text{-}62)$$

For both types of waveguide it is important to note that there is no sudden change in shielding effectiveness at the cutoff frequency. The attenuation is zero at the cutoff frequency. It increases smoothly with decreasing frequency until the term $2f/c$ or $2\pi f/c$ becomes insignificant in Equation (10-48) or (10-60). Even well below the cutoff frequency, the attenuation is proportional to the length of the waveguide. An aperture does *not* by itself behave as a waveguide, since the length of the waveguide would be very short. To achieve any attenuation greater than that due to a slot antenna, an actual waveguide must be mounted in the aperture, as shown in Figure 10-14.

We must remember that the preceding analysis is valid only for a waveguide with no conductor passing through it. It may be a hollow ventilation duct, or it may contain *fiber-optic* cables, but no conductors. A metallic cable penetrating a shield requires different treatment, which we will discuss in Chapter 11.

10-5 PROBLEMS

1. Show that with $\mathbf{J} = -\sigma\nabla V$, Equation (10-7) is a solution to Laplace's equation, $\nabla^2 V = 0$. Show that it also satisfies the boundary conditions.

2. A round ventilation hole in a shield is 12 cm in diameter. A 1000-MHz plane wave approaches the shield, with the **E** and **H** fields parallel to the plane of the shield. Consider the hole as a slot perpendicular to the **E** field. How much attenuation will this shield cause to these fields when they reach a receiving antenna located 30 m away from the shield?

3. To reduce the emissions in Problem 2, the shield is to be replaced by a perforated shield. The total hole area must be larger to provide comparable ventilation. The new shield contains 50 holes, each 3 cm in diameter. The holes are spaced such that they do not interfere with adjacent aperture currents. The same plane wave approaches the shield, and the receiving antenna is at the same location. How much attenuation does this shield now provide to the wave under the following conditions?
   (a) The relative phases of the fields are random.
   (b) All phases are equal (the worst case).

4. A panel 18 in. wide and 54 in. high is attached to a cabinet by 24 equally spaced screws around its perimeter, including one screw at each corner. The equipment in the cabinet generates a 150-MHz signal. Calculate the attenuation of this signal when it passes through the resulting slots and reaches a receiving antenna 30 m away from the shield. Assume the worst case: the phases of the signals from all slots are equal. Neglect any interference to aperture currents from adjacent slots.

5. Repeat Problem 4 for the same cabinet and equipment, but using only 8 equally spaced screws, including one at each corner.

6. Show that Equations (10-40) and (10-41) are solutions of Maxwell's equations and that they satisfy the boundary conditions for a rectangular waveguide.

7. The derivative of a Bessel function is

$$J'_n(x) = \frac{n}{x}J_n(x) - J_{n+1}(x) \tag{10-63}$$

and the divergence and curl in cylindrical coordinates are

$$\nabla \cdot \mathbf{D} = \frac{1}{\rho} \cdot \frac{\partial}{\partial \rho}(\rho D_\rho) + \frac{1}{\rho} \cdot \frac{\partial D_\phi}{\partial \phi} + \frac{\partial D_z}{\partial z} \tag{10-64}$$

$$\nabla \times \mathbf{H} = \left(\frac{1}{\rho} \cdot \frac{\partial H_z}{\partial \phi} - \frac{\partial H_\phi}{\partial z} \right) \mathbf{a}_\rho + \left(\frac{\partial H_\rho}{\partial z} - \frac{\partial H_z}{\partial \rho} \right) \mathbf{a}_\phi + \left[\frac{1}{\rho} \cdot \frac{\partial(\rho H_\phi)}{\partial \rho} - \frac{1}{\rho} \cdot \frac{\partial H_\rho}{\partial \phi} \right] \mathbf{k} \tag{10-65}$$

Show that Equations (10-55) and (10-56) are solutions to Maxwell's equations and satisfy the boundary conditions for a cylindrical waveguide.

8. How much is a 300-MHz signal attenuated by a 5-cm length of a 1-cm-by-2-cm rectangular waveguide?

9. How much is a 200-MHz signal attenuated by a 5-cm length of a cylindrical waveguide whose diameter is 2 cm?

10. A microwave transmitter operates at 1200 MHz and requires forced air cooling. The air duct is 0.25 m wide, 0.1 m high, and 0.5 m long. The 1200-MHz signal would be unacceptable if allowed to propagate through the duct unattenuated. The duct behaves as a waveguide, but its cutoff frequency is too low. It can be made higher by inserting baffles lengthwise in the duct, thus forming smaller rectangular tubes.
(a) What is the cutoff frequency?
(b) Why would a cutoff frequency of 1200 MHz still be unacceptable?
(c) How many vertical and horizontal baffles are necessary to raise the cutoff frequency to three times the operating frequency, or 3600 MHz?

11

SHIELD PENETRATION
BY WIRES AND CABLES

Our study of the effects of a hole in a shield assumed that no wires or other conductive material passed through the hole. Holes provided for ventilation, for example, would satisfy this condition. Often, however, the purpose of a hole in a shield is to allow a wire or cable to pass through it. The wire passing through the hole causes much more harm than the hole itself. We shall now study the consequences of wires or cables that penetrate a shield, and methods of dealing with them.

A hole with a conductor passing through it does not behave as a slot antenna. Instead, the wire conducts the field through the hole and then reradiates it. This process requires a completely different analysis. Neither does a waveguide with a wire inside it behave as discussed in Chapter 10. A new propagation mode, with a cutoff frequency of zero, becomes possible. Thus, the waveguide causes no attenuation and is useless.

11-1 INTERCONNECTING LEADS AS ANTENNAS

At the point where a wire or cable passes through a hole in a shield, a voltage normally exists between the wire and the shield. The voltage may contain radio frequencies generated by the equipment inside the shield. Outside the shield, the cable acts as a monopole antenna, excited by the voltage. Depending on the frequencies present in the voltage, and the resonance characteristics of the antenna, the radiated field may be as strong as with no shield at all.

If the cable passing through the shield carries radio frequencies, they may or may not be intentional. Power cables should have no RF on them, and unintentional RF is preventable by filtering. Signal cables, on the other hand, may have to carry

RF currents. As explained in Chapter 8, any RF current that returns via a nearby conductor in the same cable is differential-mode current. We noted that radiation due to differential-mode current is proportional to the area of the current loop. With closely spaced wires, this area, and therefore the radiation, is small. Radio-frequency currents might, however, flow in only one direction through the cable and return as displacement currents in space. These are common-mode currents, which are troublesome. They were discussed in Chapter 8 for unshielded sources. A cable carrying common-mode currents through a hole in a shield behaves as if a noise source were located in the hole and connected between the shield and the cable. Usually the common-mode current is unknown and must be measured. However, if a cable does not run near any metallic framework, we can estimate its common-mode current by studying the impedance of the antenna formed by the cable. Most cables, of course, are near some type of metallic framework, which changes the common-mode current on the cable. Since these effects are too complex to predict, we will assume that the cable is in free space.

The discussion of antenna impedances in Chapter 10, for slot antennas, is also applicable to a wire or cable acting as a monopole antenna. Here, however, the antenna does not receive the signal and then reradiate it. Instead, the cable is directly or capacitively connected to an RF source, which may be intentional or unintentional. The point at which it is connected is the driving point, and the ratio of the applied RF voltage to the common-mode current is the driving-point impedance Z_m. Equations (10-20) and (10-21) specify the impedance of a monopole antenna of length h and thickness w. For convenience, the equations and their graph (Figure 11-1) are repeated here.

$$R_m(h) = \frac{15\{[2 + 2\cos(2\beta h)]\,\text{Cin}(2\beta h) - \cos(2\beta h)\,\text{Cin}(4\beta h) - 2\,\sin(2\beta h)\,\text{Si}(2\beta h) + \sin(2\beta h)\,\text{Si}(4\beta h)\}}{1 - \cos(2\beta h)} \tag{11-1}$$

$$\begin{aligned}
X_m(h, w) = 15\,\Big\{ &\sin(2\beta h)\Big[-4\,\sinh^{-1}\Big(\frac{2h}{w}\Big) + 2\,\sinh^{-1}\Big(\frac{4h}{w}\Big) + \\
&2\,\text{Cin}(v_1) - 2\,\text{Cin}(u_1) + \text{Cin}(v_2) - \text{Cin}(u_2)\Big] + \\
&\cos(2\beta h)[2\,\text{Si}(u_1) - 2\,\text{Si}(\beta w/2) + 2\,\text{Si}(v_1) - \\
&\text{Si}(u_2) - \text{Si}(v_2)] + 2[\text{Si}(u_1) - 2\,\text{Si}(\beta w/2) + \text{Si}(v_1)]\Big\}
\end{aligned} \tag{11-2}$$

where

$$u_1 = \beta[\sqrt{h^2 + (w/2)^2} - h] \qquad u_2 = \beta[\sqrt{(2h)^2 + (w/2)^2} + 2h]$$

$$v_1 = \beta[\sqrt{h^2 + (w/2)^2} + h] \qquad v_2 = \beta[\sqrt{(2h)^2 + (w/2)^2} - 2h]$$

Consider, for example, a cable of length $\lambda/4$ carrying a 1-V common-mode signal. From Equations (11-1) and (11-2), the impedance Z_m is $36.5 + j21.25\ \Omega$. The common-mode current on the antenna at the driving point is

$$|I_0| = \frac{|V_0|}{|Z_m|} = \frac{1\ \text{V}}{|36.5 - j21.25|\ \Omega} = 23.7\ \text{mA} \tag{11-3}$$

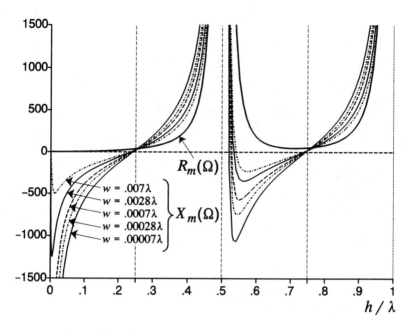

Figure 11-1 Driving-point impedance vs. monopole antenna length

Since we assume that the current distribution on a transmitting antenna is sinusoidal, the average current over the length of the cable is

$$|I| = \frac{2}{\pi}|I_0| = \frac{2V_0}{\pi|Z_m|} = 0.636 \cdot 23.7 \text{ mA} = 15.1 \text{ mA} \tag{11-4}$$

To find the radiated field due to this common-mode current, we use Equation (8-156). As before, we assume that there is a conductive floor under the cable, which could double the radiated field intensity at certain angles.

$$|\mathbf{E}_s| = \frac{f\mu_0|I|h}{r}\mathbf{a}_E = \frac{f\mu_0}{r} \cdot \frac{2|V_0|}{\pi|Z_m|} \cdot \frac{\lambda}{4}\mathbf{a}_E = \frac{\eta_0}{4r} \cdot \frac{2|V_0|}{\pi|Z_m|}\mathbf{a}_E$$

$$= \frac{94.25 \ \Omega}{r} \cdot 15.1 \text{ mA } \mathbf{a}_E = \frac{1.42 \text{ V}}{r}\mathbf{a}_E \tag{11-5}$$

At a distance of 10 m from the cable, the field strength is therefore 142 mV/m. This field is strong enough to cause severe interference to radio receivers, and it greatly exceeds the radiation limits to be discussed in Chapter 12. Yet it results from a common-mode voltage of only 1 V. It does, of course, require a cable length h equal to an odd multiple of $\lambda/4$, a resonant length. Every cable, however, will show this effect at some frequencies. If a noise source emits many frequencies, the likelihood increases that one or more cables will be resonant at a noise frequency.

Unlike a slot in a shield, the cable length usually cannot be kept shorter than $\lambda/4$. The driving-point impedance may vary from 30 Ω to several megohms, depending on the ratio h/λ. For most frequencies the impedance will be below 500 Ω, but there will always be frequencies for which the impedance is high. If a noise source contains multiple frequencies, which it usually does, then the impedance probably will be high for some frequencies and low for others. Moving a cable changes the current distribution on it and thus also changes the driving-point impedance.

We can reduce the power radiated by the cable by causing an intentional impedance mismatch between the RF source and the driving point of the cable. This is difficult to achieve over a wide frequency range, due to the large variations in driving-point impedances. Any action that reduces emissions at one frequency may increase them at other frequencies. This is particularly true when relocating cables for any reason, such as when repairing or upgrading a system. To design a reliable system that is not sensitive to cable placement, we must use an attenuation method that is independent of the driving-point impedance. Different techniques are necessary, depending on whether the cables must intentionally carry radio frequencies. These will be discussed in the following sections.

11-2 TREATMENT OF POWER AND LOW-FREQUENCY LEADS

Many cables are intended to carry only audio signals, electromechanical signals, or power distribution. Because they do not intentionally carry RF signals, they are often overlooked as potential noise radiators. They are moderately easy to deal with, but we must not ignore them. The obvious remedy is to prevent the RF signal from reaching the cable. We may shunt it to ground, or block it with series impedance, or both. Which scheme is most effective depends on the driving-point impedance Z_m of the cable at the noise frequencies.

To shunt the RF signal to ground, we use *bypass* capacitors whose impedances are small compared to Z_m. This technique is most effective for high driving-point impedances. A bypass capacitor is connected between the shield and each cable conductor at the point of penetration. The equivalent circuit appears in Figure 11-2. Impedances Z_{s1} through Z_{sn} are the impedances contained in the circuits connected to the cable leads, and depend on the type of circuit. Sources I_{s1} through I_{sn} are the RF noise sources, modeled here as current sources. The RF sources may or may not be intentional. If they are intentional then we assume that they are not intentionally connected to the cable leads but are capacitively coupled. The sources might instead be voltage sources in series with impedances, in which case Figure 11-2 depicts their Norton equivalent circuits. Impedances Z_{m1} through Z_{mn} are the driving-point impedances of the leads in the cable. For a multiconductor cable such as the one shown here, the individual driving-point impedances are higher than for a single wire. Their parallel combination is the value given by Equations (11-1) and (11-2) with w equal to the thickness of the entire cable.

Consider the example of the previous section, where a single wire was causing a field of 142 mV/m at a distance of 10 m from the cable. Assume that legal

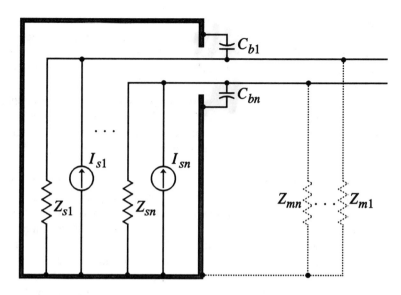

Figure 11-2 Bypass capacitors where leads penetrate cabinet

requirements require us to reduce the radiated noise field strength to 150 μV/m at that distance. Although the source impedance was not specified, we now assume it to be 50 Ω, which is typical for transistor circuits. From circuit theory,

$$V_0 = \frac{I_s Z_s Z_m}{Z_s + Z_m}$$

$$|I_s| = \left| \frac{V_0(Z_s + Z_m)}{Z_s Z_m} \right|$$

$$= \left| \frac{1 \text{ V} \cdot (50 + 36.5 + j21.25) \ \Omega}{50 \ \Omega \cdot (36.5 + j21.25) \ \Omega} \right|$$

$$= 42.2 \text{ mA} \tag{11-6}$$

To reduce the radiated field as required, V_0 must decrease by a factor of 0.15/142, to 1.05 mV. The required capacitive reactance will be much smaller than Z_s and Z_m in parallel with it, so we can ignore them. The required reactance is

$$|X_c| = \frac{1.06 \text{ mV}}{42.1 \text{ mA}} = 0.025 \ \Omega \tag{11-7}$$

At a frequency of 150 MHz, for example, this reactance would require a capacitor of 0.042 μF, or the nearest standard size, which is 0.047 μF. At higher frequencies the required capacitance would be lower.

For the more common situation of a cable containing many conductors, the conductors behave as a single antenna. The composite source impedance is the parallel combination of Z_{s1} through Z_{sn}. The driving-point impedance Z_m is the impedance of a single monopole antenna whose length and thickness are equal to the corresponding dimensions of the cable. Each conductor requires a capacitor, and their total parallel capacitance is the value calculated from the above impedances. Any lead without a capacitor would conduct the RF noise through the hole and cancel the effect of the other capacitors.

A ground wire is a special case. Ideally, the ground wire should be connected directly to the shield where it passes through. This, however, sometimes causes undesirable direct currents to flow in the shield. If a direct connection is not possible for this reason or any other reason, the ground wire must be bypassed with a capacitor, just like any other lead.

If an ideal capacitor could be made, the solution would be no more complicated than the one discussed above. However, a real capacitor contains series resistance and inductance, and shunt conductance, as discussed in Chapter 6. For the capacitor types used for bypassing, the series resistance and shunt conductance are negligible, but not the series inductance L. This inductance again causes a series resonance. Its value depends mainly on the length of the capacitor leads.

The driving-point impedance given by Equations (11-1) and (11-2) assumes that the edge of the hole in the shield fits closely around each cable conductor. A bypass capacitor connected at the point of penetration might appear as in Figure 11-3. The dotted line in the figure shows the path of the high-frequency current provided by the capacitor. The series inductance L_c can be no less than the inductance of this current loop. The inductance increases with the size of the loop, so the loop should be as small as possible. It could hardly be smaller, however, than shown in the figure.

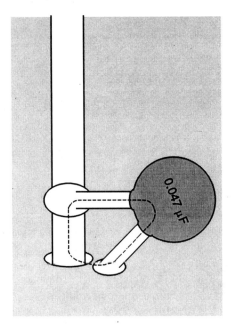

Figure 11-3   Current path at capacitor mounting point

As in Chapter 6, we estimate L_c by using the inductance formula for a circular loop. Its diameter is equal to the approximate size of the (non-circular) loop in the real circuit. For the geometry shown in the figure, typical dimensions are $r = 5$ mm and $d = 0.64$ mm (#22 AWG). These are the values assumed for the decoupling capacitor in Chapter 6, for which the approximate inductance is 0.0355 μH. With $C = 0.047$ μF, the resonant frequency is again 3.89 MHz.

As noted in Chapter 6, a capacitor is less effective above its resonant frequency. The leads in Figure 11-3 are already as short as possible, and the capacitor value is fixed by the required shunt impedance. Therefore, this geometry is not suitable when such a low impedance is required at very high frequencies such as 100 MHz. It is usable at lower frequencies or for applications where only a small capacitance is necessary, since the resonant frequency increases as the capacitance decreases.

For bypass applications requiring very low inductance, a *feed-through* capacitor is desirable. Its equivalent circuit appears in Figure 11-4. With the inductance in series with the lead penetrating the shield, as shown, it forms a low-pass filter and becomes beneficial. However, we must avoid any inductance in series with the capacitor ground, shown crossed out in the figure. This is achievable only by mounting the capacitor *in* the hole through which the lead passes. Grounding the capacitor with a wire, instead of mounting it properly in the hole, adds inductance where it cannot be tolerated. This causes the feed-through capacitor to perform as poorly as a conventional capacitor.

For a cable containing many conductors, each requires its own feed-through capacitor. To avoid a large, bulky arrangement of capacitors, filtered connectors are available. Such connectors contain a very small capacitor on each pin. Each capacitor is thus mounted in its own hole, and there is very little inductance in series with the capacitor. This arrangement is nearly ideal, but here it is even more important to insure that there is no inductance between the connector body and the shield. The connector must be mounted in the hole in the shield, and it must be well bonded to

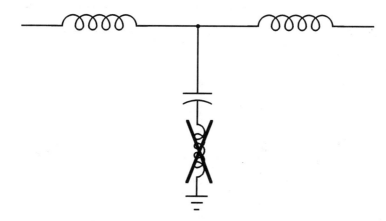

Figure 11-4   Equivalent circuit of feed-through capacitor

the shield all around the connector. A poorly mounted connector can cause capacitive coupling between leads carrying different signals, possibly causing cross talk.

Since the impedance of a bypass capacitor must be small compared to the driving-point impedance Z_m, it is useful only where Z_m is high. If Z_m is low, it is more effective to block the noise by introducing a high series impedance consisting of an inductor such as an RF choke or ferrite sleeve. The equivalent circuit appears in Figure 11-5. Here it is more convenient to model the noise as voltage sources. If they are really capacitively coupled current sources, we use their Thévenin equivalent circuits.

Consider again the example of the previous section, where it was necessary to reduce the radiated field from 142 mV/m to 150 μV/m. We again assume the source impedance to be 50 Ω. Now, from circuit theory,

$$V_0 = \frac{V_s Z_m}{Z_s + Z_m}$$

$$|V_s| = \left| \frac{V_0(Z_s + Z_m)}{Z_m} \right|$$

$$= \left| \frac{1 \text{ V} \cdot (50 + 36.5 + j21.25) \text{ } \Omega}{(36.5 + j21.25) \text{ } \Omega} \right|$$

$$= 2.11 \text{ V} \tag{11-8}$$

The common-mode RF current I_0 flowing onto the cable is

$$|I_0| = \frac{|V_0|}{|Z_m|} = \left| \frac{1}{36.5 - j21.25} \right| = 23.7 \text{ mA} \tag{11-9}$$

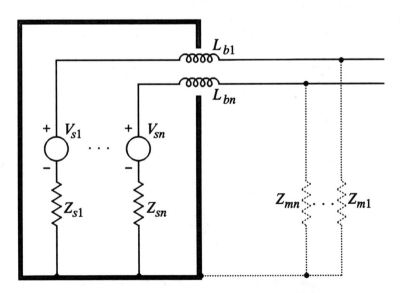

Figure 11-5   RF chokes where leads penetrate cabinet

To reduce the radiated field as required, I_0 must decrease by a factor of 0.15/142, to 25.0 microamperes. The required inductive reactance will be much greater than Z_s and Z_m in series with it, so we can ignore Z_s and Z_m here also. The required reactance is

$$|X_L| = \frac{2.11 \text{ V}}{25 \text{ } \mu\text{A}} = 84.3 \text{ k}\Omega \qquad \textbf{(11-10)}$$

At the frequency of 150 MHz assumed earlier, this reactance would require an RF choke of 89.5 μH. As with a bypass capacitor, at higher frequencies the required inductance would be less. For a cable containing multiple conductors, we generalize the analysis in the same manner as with bypass capacitors.

From Figure 11-1 we note that X_m, the reactive component of Z_m, may vary from a large negative value to $+\infty$. At a frequency for which X_m is negative, the positive reactance of a choke may cancel X_m and greatly increase the emissions. This is known as "loading" the antenna. Theoretically it can occur at any frequency for which X_m is negative. Its greatest effect is evident when $R_m \approx R_s$ and $X_m \ll -R_m$. Therefore, when we add an RF choke, we must always test its effect on emissions at lower frequencies, which might be strengthened due to loading.

Like capacitors, ideal chokes cannot be made. Practical inductors have series resistance and shunt capacitance, and it is the latter that causes most RF problems. At some frequency, the susceptance of the shunt capacitance is equal and opposite to the susceptance of the inductor itself. At this parallel-resonant frequency, the impedance is infinite and no common-mode current can flow onto the cable, which again is ideal. Above this frequency, however, the impedance decreases with frequency, and like a bypass or decoupling capacitor, the choke has the opposite of its intended effect. Thus, a choke is also less effective above its resonant frequency.

An RF choke is normally wound in a way that minimizes capacitive coupling between its turns. However, the shunt capacitance also depends on the proximity of the leads to each other. An RF voltage always exists across the choke; otherwise, the choke would be unnecessary. This implies that a moving electric flux, or displacement current, exists between the leads. This current bypasses the choke and is undesirable. To reduce it, the choke leads should be kept as far apart as possible.

A type of choke that has a particularly low shunt capacitance is a ferrite sleeve encircling the wire or cable, as shown in Figure 11-6. The only shunt capacitance for this choke is due to the electric flux which passes outside it. If the cable is straight where it passes through the sleeve, this flux is minimized. The effectiveness of a ferrite sleeve is limited only by the available inductance. We shall now show the values of inductance obtainable with such a sleeve.

If the conductor is concentric with the sleeve, the field intensity magnitude H is constant around the conductor axis for a given radius. By Ampere's law it must be the enclosed current divided by the circumference of a circle of that radius, or

$$\mathbf{H} = \frac{I\mathbf{a}_\phi}{2\pi r} \qquad \textbf{(11-11)}$$

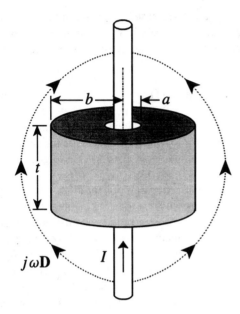

Figure 11-6 Ferrite sleeve

Note that H is the same with or without the sleeve (for the same current I), since Equation (11-11) contains no reference to μ. The flux density B, however, does increase within the ferrite material, and is

$$\mathbf{B} = \mu\mathbf{H} = \frac{\mu I \mathbf{a}_\phi}{2\pi r} \tag{11-12}$$

The *additional* flux density due to the ferrite sleeve is

$$\Delta\mathbf{B} = \mathbf{B} - \mu_0\mathbf{H} = (\mu - \mu_0)\frac{I\mathbf{a}_\phi}{2\pi r} = (\mu_r - 1)\frac{\mu_0 I \mathbf{a}_\phi}{2\pi r} \tag{11-13}$$

and the additional inductance is

$$\Delta L = \frac{\Delta\Phi}{I} = \frac{\int_S \Delta\mathbf{B} \cdot d\mathbf{S}}{I} = \frac{(\mu_r - 1)\mu_0}{2\pi} \int_0^t \int_a^b \frac{dr}{r} dz$$

$$= \frac{(\mu_r - 1)\mu_0 t}{2\pi} \ln\frac{b}{a} \tag{11-14}$$

where a, b, and t represent the dimensions shown in Figure 11-6. At very high frequencies, μ_r is seldom above 100. So, for example, the inductance of a ferrite sleeve with $\mu_r = 101$, $b/a = e$, and $t = 1$ cm is

$$\Delta L = \frac{(101 - 1) \cdot 0.4\pi \,\mu\text{H/m} \cdot 0.01 \text{ m}}{2\pi} = 0.2 \,\mu\text{H} \tag{11-15}$$

As shown earlier, a typical required value of L is 90 μH. Obviously, except for frequencies where μ_r is much higher, ferrite sleeves used in this manner cannot provide inductance values this large. Where smaller inductances are permissible, however, a ferrite sleeve provides a very high self-resonant frequency due to its low shunt capacitance.

Sometimes the displacement current is too great, even around a ferrite sleeve. The displacement current, and its associated shunt capacitance, can be reduced by using a sleeve with a larger outer radius b. This lengthens the displacement current path. Another method is to mount the sleeve in a tubular conductor, which we then mount in the hole like a feed-through capacitor. This forces all displacement current through the ferrite, since no displacement current can pass through the solid part of the shield. Like a feed-through capacitor, the tubular conductor must be well bonded to the shield.

A problem often overlooked with RF chokes is the saturation effect of direct currents. Cables protected by RF chokes often carry only low-frequency or direct currents, and these cause magnetization of the ferrite. Since ferrites are nonlinear, their permeability decreases when DC is present. Though many chokes can tolerate up to several amperes of DC, power leads often carry much higher currents. Then the inductance of the choke may decrease to a small fraction of its original value, and the choke may become ineffective. To prevent saturation, direct currents should return through another winding on the same choke or through the same sleeve. Then there is no net direct current and no saturation. If DC is unavoidable, saturation effects may be lessened by using a ferrite sleeve with a small air gap, but this greatly reduces the permeability.

For extreme cases, capacitors and chokes may be used together. Filters containing tee, pi, or ladder networks are commercially available. Their specifications normally state their corner frequencies and attenuation per decade. It is important to note that these specifications normally assume a fixed termination impedance that is often not realistic. The actual termination impedance will normally vary over a wide range, as shown in Figure 11-1. A filter designed for a termination impedance of 50 Ω may perform completely differently when terminated by an arbitrary impedance. Inductance in the ground connection also impairs the performance of an RF filter. It must be mounted through the hole in the shield through which the lead passes, like a feed-through capacitor.

A capacitor or choke diverts or blocks the RF emissions, but the RF energy still exists within the shielded enclosure. It is better still if we can dissipate the RF energy, since then the energy is less likely to couple to other leads. To dissipate the energy requires a device that exhibits losses of some type. An example is a coil core or sleeve made of a magnetic material of moderate conductivity. Eddy currents then flow in the core or sleeve, as in a transformer secondary, and these currents dissipate RF energy as heat in the material. Some cores exhibit hysteresis losses, which also dissipate RF energy. These devices work best at very high frequencies and often provide better attenuation than lossless inductors or capacitors.

11-3 TREATMENT OF HIGH-FREQUENCY LEADS

For cables that must intentionally carry radio frequencies, treatment is necessarily more complicated. Obviously we cannot simply bypass or block all RF signals, for then the cable could no longer perform its intended function. As before, we must distinguish the desired signals from noise, but we cannot do so by frequency alone.

As noted earlier, for cables of length comparable to λ, most radiation is due to common-mode currents. Differential-mode currents cause little radiation if the area enclosed by the current loop is small. Therefore, if we transmit the intentional RF signals in the differential mode and prevent common-mode currents, we reduce unwanted RF emissions. Similarly, if a receiver is sensitive only to differential-mode signals, it is more immune to RF noise. We shall now discuss methods to reduce common-mode currents in external cables. This topic was discussed in Chapter 4 with regard to inductive coupling between cables. Most of this material also applies to radiation from cables.

As before, to reduce the common-mode current, we must increase the common-mode impedance. We can achieve this by adding series inductance, as discussed in the previous section. In cables that intentionally carry RF signals, however, we must not change the differential-mode impedance, or we will distort the desired signal. We can achieve this by inserting a common-mode choke into the leads carrying the differential-mode signal. It functions as explained in Section 4-6, except that the common-mode current return path passes through space instead of through a common ground. The common-mode inductance of each winding of the choke is

$$L = (1 + k)L_c \qquad \textbf{(11-16)}$$

where L_c is the inductance of either winding and k is the coupling coefficient, approximately equal to 1. The necessary common-mode inductance is calculated as though each winding were a separate inductor. Its impedance must be high compared to Z_m, which is given by Equations (11-1) and (11-2). The differential-mode inductance introduced into the signal path is

$$L_d = 2(1 - k)L_c \qquad \textbf{(11-17)}$$

This should be as small as possible, which it will be if $k \approx 1$. Its actual value should be used in any calculations involving the signal current loop.

The best method of reducing common-mode currents is to use a shielded cable. If the cable shield is bonded to the cabinet shield all around the hole, all RF fields will remain inside the cable shield and will not radiate. This usually requires a coaxial connector mounted in the hole. Since RF energy is intentionally conveyed to the other end of the cable, the circuitry there may also require a shielded enclosure. If it does, the cable shield also should be bonded to that cabinet. Then all RF fields are within the cable shield and the two cabinets, which minimizes radiation. If the circuitry at the opposite end is not shielded, or if the cable shield is not bonded to that cabinet, RF radiation will escape at the open end of the cable.

Radio-frequency currents can then flow on the outside of the shield, which then acts as an antenna driven at the far end. Radiation may be as severe as it would be with no shield at all. Currents on the outside of the shield are common-mode currents. Thus, they can be reduced by routing the shielded cable through a ferrite sleeve, which then acts as a common-mode choke.

Complex systems may require many shielded cables between two or more cabinets. Then, to keep the areas of the differential current loops small, we must force each RF current to return via its own shield. A ferrite sleeve or other common-mode choke on each lead helps assure this by increasing the impedances of the undesired current paths. To minimize the current loop areas, the shielded cables should be bundled together tightly. Preferably, all shielded cables should be in the same cable jacket.

11-4 PROBLEMS

1. A 5-conductor cable protrudes a distance of 1.8 m after passing through a hole in a shield, and its thickness is 8.4 mm. It is meant to carry only direct current, but a signal of 100 MHz is somehow getting onto it. The RF voltage measured between the conductors and the shield is 0.1 V. The cable does not pass near any metal or other conductive material, except where it passes through the shield. What is the radiated field intensity at a distance of 10 m from the cable?

2. In Problem 1, each conductor in the cable is connected to a source or load having an impedance of 100 Ω. What size of capacitor is required on each conductor to reduce the radiated field to 150 μV/m at a distance of 10 m from the cable?

3. In Problems 1 and 2, a choke is to be used instead of a capacitor to reduce RF emissions. What size of choke is required on each conductor to achieve the same reduction as that provided by the capacitor?

4. To cure an emission problem at 100 MHz, a 5-μH choke is to be installed on a thin cable that is 4.5 m long. However, the noise source also emits at 10 MHz. At this frequency, $Z_m = 10 - j313$ Ω at the driving point of the cable. Assume the noise source impedance is 10 Ω.

   (a) Do the emissions at 10 MHz increase or decrease?

   (b) By what ratio does the radiated field intensity at 10 MHz change?

12

EMC REGULATIONS
AND MEASUREMENTS

Radiated electromagnetic interference can travel for long distances, and it knows no boundaries. It can affect equipment owned and used by someone other than the owner of the interfering device. The interfering device may function perfectly well while rendering someone else's appliance useless. Without regulations, the affected user would have no recourse.

In the United States, the governing body that controls electromagnetic emissions by civilian equipment is the Federal Communications Commission (FCC). Similar agencies exist in many other countries. Military and other government equipment is governed, not by the FCC, but by its own set of standards. These have been in force for a longer time, and differ considerably from the civilian regulations.

12-1 CIVILIAN REGULATIONS

Regulations formulated and enforced by the FCC are intended to guarantee that a device is not likely to interfere with a nearby receiver. Equipment must conform to these regulations before it can be advertised or sold commercially. The manufacturer must design his product with these regulations in mind and must test a prototype to assure that his design is acceptable. For some types of equipment, he must apply to the FCC for certification, submitting his test results with the application. Otherwise, he must keep his test results on file, subject to inspection by the FCC.

Any unintentional radiating device is regulated by the FCC.[1] These rules distinguish between two types of devices. *Class-A* devices are those intended to be used commercially, while *Class-B* devices are those intended for use in the home. Since Class-B devices are more likely to be placed near someone else's television set, for example, the rules governing Class-B devices are more stringent. They follow from the reasoning below.

The rules assume that a Class-B device is not likely to be located within 10 m of a television receiving antenna. If located closer probably it could be moved. Therefore, the radiation limits should be such that a radiating device located 10 m from a TV receiving antenna will cause no detectable interference.

The geographical area of coverage of a TV broadcast station is defined in terms of its received field strength. In the coverage area where the signal is of Grade A, a television signal on the low band (channels 2 through 6, which cover 56–88 MHz) must have a field strength of at least 2250 μV/m, or 67 dB$_{\mu V/m}$. An unintentionally radiating device must not interfere with a TV set receiving a TV signal of this strength.

Specifications state that a television receiver must tolerate any signal-to-noise ratio greater than 180, or 45 dB. Thus, if all noise is at least 45 dB weaker than the TV signal strength, the noise is not detectable on a properly operating TV receiver. The maximum noise level that would not interfere with the weakest Grade-A television signal received by the "worst" TV receiver that is still within specs is therefore

$$E_n = \frac{2250\ \mu\text{V/m}}{180} = 12.5\ \mu\text{V/m} \qquad \textbf{(12-1)}$$

or

$$(E_n)_{\text{dB}\mu V/m} = 67\ \text{dB}_{\mu V/m} - 45\ \text{dB} = 22\ \text{dB}_{\mu V/m} \qquad \textbf{(12-2)}$$

To assure no interference to any properly operating TV set receiving a Grade-A signal, the maximum noise field strength allowed to reach its antenna must not exceed this value.

Usually at least one wall will be located between the radiating device and the antenna. This wall is likely to attenuate the noise by a factor of 2.5, or 8 dB. The rules assume that the wall will be present during normal operation but not during testing. So if the emissions are measured *without* a wall at a distance of 10 m from the radiating device, they must not exceed 2.5 times the above value, or

$$E_n = 12.5\ \mu\text{V/m} \cdot 2.5 = 31.25\ \mu\text{V/m} \approx 30\ \mu\text{V/m} \qquad \textbf{(12-3)}$$

or

$$(E_n)_{\text{dB}\mu V/m} = 22\ \text{dB}_{\mu V/m} + 8\ \text{dB} = 30\ \text{dB}_{\mu V/m} \qquad \textbf{(12-4)}$$

1. "Part 15—Radio Frequency Devices," in *Code of Federal Regulations, Volume 47: Telecommunication* (Washington, D.C.: U.S. Government Printing Office, 1991), subpt. A–B.

IF 1 mW gives 0·5 V/m
then 36 mW gives 3 V/m at same distance

For convenience, measurements of Class-B devices are made at 3 m instead of 10. The FCC regulations for radiated emissions apply only to frequencies above 30 MHz, for which $\lambda = 10$ m and $\lambda/(2\pi) \approx 1.67$ m. Therefore, at either 3 m or 10 m the measurements are made in the far-field region, and the field strength is inversely proportional to the distance. Thus, to assure a field strength no greater than 30 μV/m at a distance of 10 m, the limit at 3 m must be

$$E_n = \frac{10}{3} \cdot 30 \ \mu V/m = 100 \ \mu V/m \qquad (12\text{-}5)$$

or

$$(E_n)_{\text{dB}\mu V/m} = 30 \ \text{dB}_{\mu V/m} + 10 \ \text{dB} = 40 \ \text{dB}_{\mu V/m} \qquad (12\text{-}6)$$

This is the Class-B limit for this frequency range.

Class-A devices are usually larger than Class-B devices and are less likely to be located near a TV receiving antenna. The regulations are intended to assure that a Class-A radiating device located 30 m (instead of 10 m) from a TV antenna will cause no interference to a Grade-A TV signal. As with Class-B devices, the rules assume that there will be at least one wall between the radiating device and the antenna. From Equations (12-3) and (12-4), the maximum permissible noise field strength at a distance of 30 m, without a wall, is 30 μV/m or 30 dB$_{\mu V/m}$. Due to the size of most Class-A equipment, measurements must be made at a distance of 10 m instead of 3. To assure a field strength no greater than 30 μV/m at a distance of 30 m the limit at 10 m must be

$$E_n = \frac{30}{10} \cdot 30 \ \mu V/m = 90 \ \mu V/m \qquad (12\text{-}7)$$

or

$$(E_n)_{\text{dB}\mu V/m} = 30 \ \text{dB}_{\mu V/m} + 9 \ \text{dB} = 39 \ \text{dB}_{\mu V/m} \qquad (12\text{-}8)$$

This is the Class-A limit for this frequency range, which is three times the Class-B limit normalized to the same distance.

Limits for other frequency ranges depend on the types of radio services that use the respective frequencies. Television channels in the VHF high band (channels 7 through 13) and the UHF band use stronger signals, so the noise limits are higher for these frequencies. Receivers that do not use video signals can usually tolerate a signal-to-noise ratio of 30, or 30 dB, instead of the 45 dB required by TV receivers. Thus, the noise limits are also higher for non-video services. To conform with international regulations, the Class-A limits are not always exactly three times the Class-B limits, but are roughly so. The limits for all frequencies are given in Table 12-1.

Many ambient conditions can affect the emissions from an unintentional radiating device. A device that complies with the applicable regulations may nevertheless interfere with a receiver. If it does, then the user of the interfering device must do whatever is necessary to reduce or eliminate the interference. The FCC

TABLE 12-1   FCC Limits for Radiated Emissions

Frequency	Class A limits at 10 m	Class B limits at 3 m
30–88 MHz	90 μV/m	100 μV/m
88–216 MHz	150 μV/m	150 μV/m
216–960 MHz	210 μV/m	200 μV/m
Above 960 MHz	300 μV/m	500 μV/m

regulations also require certain user instructions to be furnished with a device that may unintentionally radiate. The instructions must contain suggested remedies for harmful interference.

12-2 UNITED STATES MILITARY STANDARDS

Equipment owned and operated by the United States government is not subject to FCC regulations or any others concerning electromagnetic radiation. Every government agency that uses electronic equipment must specify, as a condition of purchase, the radiation limits that the equipment must meet. To maintain standardization, there is a common set of standards that government agencies may quote in their specifications. This is MIL-STD-461, which contains radiation limits for many different types of equipment. At the time of publication it is under extensive revision, so we will discuss it only superficially.

Whereas the FCC regulations for radiated emissions apply only to frequencies above 30 MHz, military standards extend to much lower frequencies. At lower frequencies, measurements have to be made in the near field, where the ratio of |**E**| to |**H**| is not constant. Therefore, military standards specify limits for the magnetic field as well as the electric. Military specifications also include limits on susceptibility to external fields, which are not currently included in the FCC regulations. Other differences will become evident when we discuss measurement procedures in the next section.

12-3 MEASUREMENT OF RADIATED EMISSIONS[2]

Radiated emissions depend heavily on any conductive material surrounding the source. To obtain repeatable data, the same ambient conditions must be used for all tests. Ideally, equipment should be tested in free space, far removed from any

2. "Methods of Measurement of Radio-Noise Emissions from Low-Voltage Electrical and Electronic Equipment (ANSI C63.4–1988)," *American National Standards Institute (ANSI) Standards* (New York, N.Y.: The Institute of Electrical and Electronics Engineers, Inc., 1989).

conductive, dielectric, or magnetic material. This, of course, is impossible, and the best test conditions obtainable on earth are in a large open outdoor area with very low ambient electromagnetic radiation. This is called an *open-field site*. Even at such a site, the conductivity of the soil below it will affect the fields, and it may vary at different sites or different times. To eliminate this variable, we place a large conductive floor, such as steel mesh wire,[3] on the surface of the soil. In normal use of an interfering device, a highly conductive floor, such as one made of metal, can cause the emissions from a radiating device to double at certain angles, as shown in Chapter 8. The conductive floor used for radiation testing thus causes the results to be worst-case, which is the safest test procedure.

Radiated emissions must be measured in all directions from the radiating device. The easiest way to accomplish this is to rotate the device, rather than moving the antenna. Therefore, we normally place the device on a rotating table or platform and rotate it while testing.

Ideally, the electromagnetic field should be measured using a spectrum analyzer or EMI receiver connected to a dipole antenna. Then the input voltage to the receiver can be easily related to the received field strength. Normally the dipole antenna consists of two telescoping monopole whip antennas, with the feed line connected between them. We noted in Section 10-2 that the driving-point impedance Z_m of a monopole antenna becomes purely real at a length slightly shorter than $\lambda/4$. There we were only concerned with approximate impedances. For the measurements discussed here, however, the impedance match between the antenna and receiver becomes very critical. Any mismatch between the antenna and receiver causes reradiation, which affects our calculations. Therefore it is important to use an antenna for which the driving-point impedance Z_m is purely real. The formulae of Section 10-2 apply as before, but we must now use them with much greater accuracy. The graphs of R_m and X_m appear in Figure 12-1, with scales expanded around the lengths for which X_m is zero.

The exact resonant length of a monopole whip antenna depends on its thickness w. To provide sufficient mechanical strength, the thickness of a telescoping whip antenna must usually be at least 0.01 times its length, or 0.0025λ for a quarter-wave monopole. From Figure 12-1, for $w = 0.0028\lambda$ the resonant length is 0.238λ. At this length, Z_m is purely real and equal to 32 ohms. A resonant dipole antenna would be double this length, or 0.476λ, and the impedance at its center is the impedance of two monopoles in series, or 64 ohms.

To measure the field strength that the antenna receives, we must relate this field strength to the antenna terminal voltage. First we short-circuit the dipole antenna at its center and immerse it in an electromagnetic field of intensity **E**. From Equation (10-24) with $x_1 = 0$, the current I_0 at its center is

$$I_0 = \frac{|\mathbf{E}|_s \lambda}{2\pi Z_m(l/2)} \tan\left(\frac{\pi l}{2\lambda}\right) \tag{12-9}$$

3. In the FCC measurement procedures, the mesh screen is called a ground plane. To distinguish it from a ground plane on a printed wiring card, we will refer to it as a conductive floor.

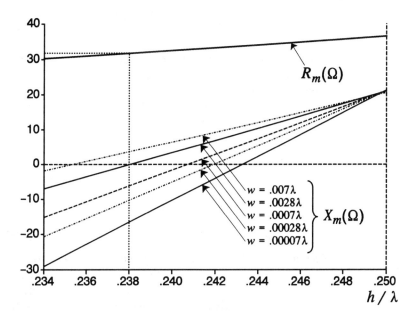

Figure 12-1 Driving-point impedance vs. monopole antenna length

where l is the dipole length and $Z_m(l/2)$ is the driving-point impedance of either half of the dipole antenna. For the resonant antenna discussed above, $l = 0.476\lambda$, $Z_m(l/2) = 32$ ohms, and

$$I_0 = \frac{|\mathbf{E}|_s \,\lambda \tan(0.238\pi)}{2\pi Z_m(l/2)} \tag{12-10}$$

We may consider the current I_0 at the center of the dipole as if it were generated by a source whose series impedance is $2Z_m(l/2)$. We now open the dipole at its center and connect the transmission line, terminated by the receiver input. Its impedance must be equal to $2Z_m(l/2)$, or 64 ohms; if not, we must use a matching circuit and calibrate the receiver accordingly. This input impedance adds to the source impedance of $2Z_m(l/2)$, and the total impedance is doubled; therefore, the current is reduced to $I_0/2$. The voltage V at the antenna terminals is

$$V = \frac{I_0}{2} \cdot 2Z_m(l/2) = \frac{|\mathbf{E}|_s \,\lambda \tan(0.238\pi)}{2\pi Z_m(l/2)} Z_m(l/2)$$

$$= \frac{|\mathbf{E}|_s \,\lambda \tan(0.238\pi)}{2\pi} = 0.148\,|\mathbf{E}|_s \,\lambda \tag{12-11}$$

The ratio of V to $|\mathbf{E}|_s$ is known as the *effective length* of the antenna, designated l_{eff}. For the resonant half-wave dipole discussed here, it is

$$l_{eff} = \frac{V}{|E|_s} = 0.148\lambda = 0.148\frac{c}{f} = \frac{44.4 \text{ m/}\mu s}{f} \qquad (12\text{-}12)$$

This is roughly one third of the physical antenna length. For a uniform field $\mathbf{E}$, the effective length is the distance between two points in space, in the direction of $\mathbf{E}$, that differ by the potential V. It thus relates the field strength $|E|_s$ to the antenna terminal voltage V when the antenna is parallel to $\mathbf{E}$. The field strength is expressible as $dB_{\mu V/m}$, in terms of $V_{dB\mu V}$, as follows:

$$E_{dB\mu V/m} = 20 \log \frac{|E|_s}{1 \ \mu V/m} = 20 \log \frac{V}{1 \ \mu V \cdot l_{eff}/1 \text{ m}}$$

$$= 20 \log \frac{V}{1 \ \mu V} - 20 \log \frac{l_{eff}}{1 \text{ m}}$$

$$= V_{dB\mu V} - 20 \log l_{eff} \qquad (12\text{-}13)$$

with l_{eff} in meters. The quantity $(-20 \log l_{eff})$ is the effective length expressed in negative dB, commonly called the *antenna factor*. From Equation (12-13) we note that it is the number of decibels that we must add to the antenna terminal voltage to calculate the field strength. For the half-wave dipole, the antenna factor AF is, from Equation (12-12),

$$AF = -20 \log l_{eff} = -20 \log \frac{44.4 \text{ m/}\mu s}{f}$$

$$= 20 \log f - 32.9 \ dB_{\mu V/m} \qquad (12\text{-}14)$$

The above formulae are valid only if the antenna is resonant, which occurs only if it is the exact length for which $X_m = 0$, or Z_m is purely real. Then, if the receiver input impedance and/or the characteristic impedance of the feed line are not equal to $2R_m$, a matching network is necessary. Any mismatch invalidates the calculations. At nonresonant lengths, the driving-point impedance of a dipole antenna varies greatly, as observed in Section 10-2, and a suitable match becomes impossible. Then the antenna factor, which depends on the driving-point impedance, also varies greatly. To obtain accurate measurements, we must adjust the length of a half-wave dipole for each frequency being measured. Thus, a half-wave dipole antenna is not usable for high-speed scanning over a band of frequencies.

Other antenna types provide a more constant impedance. An example of such a broadband antenna is the biconical antenna, shown in Figure 12-2. This antenna covers the frequency range of 20 MHz through 300 MHz and requires no adjusting. Its driving-point impedance is more constant over frequency than that of a fixed-length dipole. Nevertheless, the voltage standing-wave ratio (vswr) may approach 12:1 at some frequencies in the above range. Of course, none of the calculations used for the dipole are applicable to the biconical. The antenna factor of a biconical is normally specified by its manufacturer, who measures it and supplies a graph with the antenna. It is not very constant or linear with frequency. It assumes a fixed receiver input impedance, and attempts to account for the mismatch between this impedance and the antenna driving-point impedance. However, the mismatch

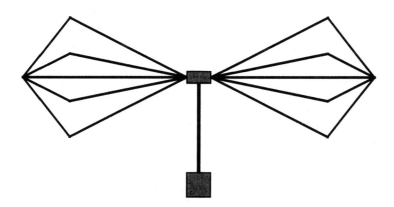

Figure 12-2 Biconical antenna

causes the antenna to reradiate, which can affect the antenna factor unless the open-field site is nearly perfect. If the receiver input impedance is not exactly as specified or not purely resistive, this compounds the problem. This, however, is correctable by using a resistive attenuation pad at the receiver input. An attenuation pad of 6 to 10 dB usually corrects the problem if the receiver is sensitive enough to make up for the attenuation.

Normally a biconical antenna should only be used for rough measurements, to search for emissions that might be troublesome. For this type of testing, high-speed frequency scanning is obviously desirable, and the biconical antenna makes it possible. For a more precise analysis at a troublesome frequency, a resonant dipole is more accurate and is preferable. At any rate, the FCC regulations assume a half-wave dipole antenna. If any other antenna such as a biconical is used, the user must demonstrate that the results are equivalent.

Dipole and biconical antennas are *balanced*, which means that they expect a high impedance between either antenna terminal and ground. Compare this with the balanced circuits discussed in Chapter 4. A coaxial cable is not balanced, and if it is connected directly to the antenna it will upset the balance. Common-mode currents will then flow on the coaxial cable and reduce the measurement accuracy. To prevent this, a *bal*anced-to-*un*balanced (*balun*) circuit, resembling a common-mode choke, is built into the antenna at the point where the coaxial cable is connected. The balun must be built with more precision than the common-mode chokes discussed in Chapter 4 in order to minimize losses. Any losses in the balun (typically 0.5 dB) must be added to the antenna factor.

An open-field test site[4] appears in Figure 12-3. The site requires an open area of dimensions $R\sqrt{3}$ by $2R$, where R is the measurement distance, normally 3 m for Class-B measurements and 10 m for Class-A. The area within the ellipse must be free of trees and other obstructions, since these could affect the measurements. The

---

4. "Guide for Construction of Open Area Test Sites for Performing Radiated Emission Measurements (ANSI C63.7–1988)," *American National Standards Institute (ANSI) Standards* (New York, N.Y.: The Institute of Electrical and Electronics Engineers, Inc., 1988).

soil within this area is covered with the conductive floor material mentioned above, usually mesh wire. We place the equipment under test (*EUT*) on a nonconductive rotating platform at one focal point of the ellipse. If the EUT is a device normally used on a table, such as a personal computer, we place it on a nonconductive table 1 m high. The receiving antenna is on a nonconductive mast or tripod, at the other focal point of the ellipse. Its height must be variable from 1 to 4 m to search for the maximum vertical angle of radiation as required by the regulations. The EUT is tested with all normal cables, including its power cord, in their typical positions. The receiver antenna cable, and any power cables leaving the test area, are under the conductive floor so as not to disturb the electromagnetic fields being measured.

The antenna factor assumes that the antenna is parallel to **E**. To find **E** without advance knowledge of its direction would require three measurements. In a spherical coordinate system with the origin at the EUT and with the z axis vertical, we would perform measurements with the antenna respectively parallel to $\mathbf{a}_r$, $\mathbf{a}_\theta$, and $\mathbf{a}_\phi$. We would thus measure E_r, E_θ, and E_ϕ, which together give **E**. However, the measurements are in the far-field region. From Equations (8-37) and (8-84) of Chapter 8 we note that regardless of the orientation of the electric dipoles and the current loops in the EUT, there can be no r component of **E** in the far-field region. Thus we need not measure E_r but must measure E_θ and E_ϕ. These are commonly known, respectively, as the *vertically* and *horizontally polarized* components of **E**. Two measurements are thus necessary, one with the antenna vertical (aligned with $\mathbf{a}_\theta$) and the other with it horizontal (aligned with $\mathbf{a}_\phi$). The total field **E** is the vector sum of these results, E_θ and E_ϕ. Both, however, are complex phasors, and they may not be in phase. If not, then the wave is elliptically polarized, and computation of $\|\mathbf{E}\|$ would require

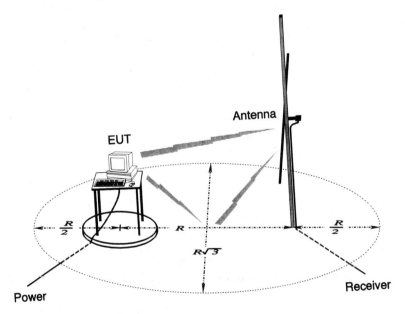

Figure 12-3   Plan of an open-field test site

knowledge of their relative phases. To avoid this difficulty, the FCC procedures specify that the value used for $\|E\|$ should be the greater of E_θ and E_ϕ. This can be no less than $\|E\|/\sqrt{2}$, which the FCC considers an acceptable approximation.

To perform the measurements, we operate the EUT in its normal manner. While slowly rotating the EUT on the platform, we vary the height of the receiving antenna above the floor. For vertical polarization, $\mathbf{a}_\theta$ changes direction as the antenna is raised, which implies that we would have to tilt the antenna to keep it parallel with $\mathbf{a}_\theta$. However, the FCC procedure states that this is not necessary, so we keep the antenna vertical and ignore the error. We record the maximum field strength measured at each frequency and each polarization. Normally, we use a broadband antenna, such as a biconical, for initial measurements. If any radiated emissions near the FCC limits are evident, we retest only those frequencies with a resonant dipole antenna.

12-4 TEST-SITE CALIBRATION

The open-field site used to make FCC measurements can significantly affect the results. Therefore, before any site can be used, it must be checked with a *known* source. Normally, we use a transmitter, driving another half-wave dipole antenna, in place of the EUT in Figure 12-3. Testing standards[5] prescribed by the FCC dictate that the transmitting antenna must be mounted at a height of 2 m above the conductive floor. The field is picked up by the half-wave dipole antenna driving the receiver, and the ratio of the transmitted voltage to the received voltage is computed. This ratio, expressed in decibels, is known as the *site attenuation*, since it is the amount by which the test site attenuates the signal between the two antenna terminal pairs.

Like the receiving antenna, the transmitting antenna must be resonant so that its driving-point impedance is purely real. Thus, if its thickness w is 0.0028λ, its length l must be 0.476λ, and the driving-point impedance $2Z_m = 64$ ohms. If the antenna is driven by a generator of voltage V_0, then the driving-point current I_0 is $V_0/(2Z_m)$. Since the current on a center-driven half-wave dipole is sinusoidally distributed, the average current I over the antenna length is equal to $2I_0/\pi$, or

$$I = \frac{2}{\pi}I_0 = \frac{2}{\pi}\frac{V_0}{2Z_m(l/2)} = \frac{V_0}{100.5\ \Omega} \tag{12-15}$$

From Equation (8-156), the maximum field intensity at a distance r from a current source of length l carrying an average current I is

$$|\mathbf{E}_s| = \frac{f\mu_0|I|l}{r}\mathbf{a}_E = \frac{1.26\ \mu\text{H/m}\cdot f|I|l}{r}\mathbf{a}_E \tag{12-16}$$

The actual field intensity varies between zero and this maximum value, depending on the height above the conductive floor, and assuming 100-percent reflection from it. This is the worst case, which should be used for design purposes. For the

---

5. "Measurement of Radio-Noise Emissions (ANSI C63.4–1988)," *ANSI Standards,* pp. 12–16.

site-attenuation calculations discussed here, however, we require a more typical figure. The FCC measurement procedures assume a reflection coefficient of 64 percent, or +4.3 dB, from the conductive floor. Therefore, the field intensity at a maximum point is 1.64 (instead of 2) times its free-space value, or

$$|\mathbf{E}_s| = 1.64 \frac{f\mu_0|I|l}{2r} \mathbf{a}_E = \frac{1.03 \ \mu\text{H/m} \cdot f|I|l}{r} \mathbf{a}_E \tag{12-17}$$

Since $l = 0.476\lambda$, $fl = 0.476f\lambda = 0.476c$, and

$$|\mathbf{E}_s| = 1.64 \frac{0.476c\mu_0|I|}{2r} \mathbf{a}_E = \frac{1.47.1 \ \Omega \cdot |I|}{r} \mathbf{a}_E \tag{12-18}$$

The FCC procedures also assume a loss of 6-percent, or 0.5 dB, in the balun built into the antenna. Let V_G represent the generator voltage before this loss, so that $V_0 = (1 - 0.06)V_G$. The field intensity at a distance r from the transmitting antenna, and at a height that produces the maximum intensity, is

$$|\mathbf{E}_s| = \frac{147.1 \ \Omega}{r} \cdot \frac{|V_0|}{100.5 \ \Omega} \mathbf{a}_E = \frac{1.463|V_0|}{r} \mathbf{a}_E$$

$$= \frac{1.463(1 - .06)|V_G|}{r} \mathbf{a}_E = \frac{1.376|V_G|}{r} \mathbf{a}_E \tag{12-19}$$

The receiving antenna is at the distance R from the transmitting antenna. We vary its height above the conductive floor until we obtain the maximum field intensity. This antenna is also resonant and is matched to the receiver. From Equation (12-11), and setting $r = R$, the voltage magnitude $|V_A|$ at its terminals is

$$|V_A| = 0.148 \ \|\mathbf{E}\| \ \lambda = 0.148\lambda \frac{1.376|V_G|}{r}$$

$$= 0.148 \cdot \frac{c}{f} \cdot \frac{1.376|V_G|}{R} = \frac{61.1 \ \text{m/}\mu\text{s}}{fR} |V_G| \tag{12-20}$$

The receiver antenna also contains a balun, and the FCC procedures assume a 6-percent loss here also. The voltage magnitude $|V_R|$ at the receiver input is therefore

$$|V_R| = (1 - 0.06)|V_A| = \frac{57.4 \ \text{m/}\mu\text{s}}{fR} |V_G| \tag{12-21}$$

The site attenuation A is the ratio of $|V_G|$ to $|V_R|$, expressed in decibels:

$$A = 20 \log \frac{|V_G|}{|V_R|} = 20 \log \frac{fR}{57.4 \ \text{m/}\mu\text{s}}$$

$$= 20 \log f + 20 \log R - 20 \log 57.4$$

$$= 20 \log f + 20 \log R - 35.2 \ \text{dB} \tag{12-22}$$

with f in megahertz and r in meters.

The FCC specifies that the site attenuation must be checked using Equation (12-22) for frequencies above 70 MHz. Below this frequency, at a distance R of 3 m, near-field effects cause the formula to become inaccurate. The value of A for f = 70 MHz and R = 3 m is 11 dB. Empirical data shows that A remains fairly constant at frequencies between 30 and 70 MHz; its value is approximately 11 dB.

The FCC requires site-attenuation measurements for vertical and horizontal polarizations. We assumed horizontal polarization in the above discussion, but vertical polarization behaves similarly. The proximity of the conductive floor to the antennas affects the antenna factor at low frequencies, however, and complicates the calculations. We will not derive the site-attenuation formula for vertical polarization, but tables are available.[6]

If the measured site attenuation is within 3 dB of the theoretical value specified by the FCC, which is the value given by Equation (12-22) at frequencies above 70 MHz, then the site is considered acceptable. Site-attenuation data is then normally placed on file with the FCC.

Although an open-field site yields the most consistent data possible, it is inconvenient to use. A true open-field site has no enclosure of any kind and is at the mercy of the weather. Sites have been enclosed with nonconductive material such as canvas or air-supported plastic, but even this affects measurements when it becomes wet due to rain. For these reasons, indoor sites are usually used for preliminary measurements. It must then be understood that the results are only approximate due to the surrounding structure. If the EUT appears to be well within the limits, outdoor testing may not be necessary.

Ambient electromagnetic noise also presents problems at most outdoor and indoor sites. Such noise is due to licensed radio stations, nearby computing equipment, and other sources. The only place to avoid this interference is inside a shielded room. Shielded rooms are prescribed for MIL-STD-461 testing. However, they are unacceptable for FCC radiation testing, because these regulations are designed for open-field testing. Results are greatly affected by internal reflections from the shield. Measurements made inside a normal shielded room may be in error by over 40 dB, or a factor of several hundred, due to "hot spots" in the room. Hot spots can be reduced by placing a lossy ferrite block, about the size of a brick, at the point of maximum **H** field strength. This absorbs some energy and reduces the error, but not enough to emulate an open-field site.

Special shielded rooms, called anechoic chambers, have their walls lined with a lossy material. If the shield impedance Z_s of this material (see Section 9-2) could be made equal to η_0, there would be no reflections and the room would behave like an open field. A perfect match is nearly impossible over all frequencies, however. The lossy material is formed into large pyramids, with the intent of causing multiple reflections and attenuating each reflection. For this to occur, the dimensions of the pyramids must be comparable to a wavelength, which is obviously impossible at frequencies near 30 MHz, where λ = 10 m.

---

6. Ibid.

To use an anechoic chamber for FCC testing, the site attenuation must be measured as for a true open field. If the measured site attenuation is within 3 dB of the theoretical value for an open field, at all frequencies for which the chamber is to be used, then it is usable for FCC testing. This is an ideal test environment, but it is very difficult to achieve, particularly at frequencies as low as 30 MHz.

12-5 MEASUREMENT OF CONDUCTED EMISSIONS

In Chapter 1 it was mentioned that an electronic device also can emit high-frequency noise on its AC power cord. The power line may then conduct this noise to another device that shares it. This book addresses radiated interference, but for completeness we will briefly discuss conducted interference.

Conducted interference is preventable by filtering the power line. Filtering techniques are presented in many textbooks and will not be discussed here. We must note, however, that high-frequency filters present the same problems as bypass capacitors and RF chokes, discussed in Chapter 11. The designer must minimize any inductance in series with a capacitive filter, by mounting the filter in the hole through which the power lead passes. Similarly, displacement current must not be allowed to bypass an inductive filter, which also should be mounted through the hole.

For Class-A devices, FCC rules restrict conducted emissions to 3000 μV at frequencies from 1.705 to 30 MHz. At frequencies from 0.45 to 1.705 MHz, the limit is 1000 μV to protect the AM broadcast band. For Class-B devices, the limit is 250 μV at all frequencies from 0.45 to 30 MHz. The voltage injected onto the AC line is measured with a spectrum analyzer or radio noise receiver, using an RF probe instead of the antenna used for radiated measurements. To assure consistent data, we use a special *Line-Impedance Stabilization Network* (*LISN*) to control the driving-point impedance of the AC line. To reduce ambient noise, conducted interference testing is normally performed in a shielded room, which is acceptable since reflections are not a problem.

12-6 PROBLEMS

1. In a certain personal computer, the fundamental clock frequency f_0 is 29 MHz with a 50-percent duty cycle, and the logic rise time t_r is 5 ns. The peak current on each logic input is 0.2 mA, the average area a of all clock current loops is 10 cm^2 per loop, and up to 100 clock inputs may switch at once. Find the frequency at which the clock circuitry is expected to cause this personal computer to fail to meet the appropriate FCC limits for radiated emissions. Use the techniques of Chapter 8, and neglect the errors inherent in EMI predictions. Prove your answer using a plot of the expected field strength versus frequency, superimposing the FCC limits on your plot.

2. Compute and plot the site attenuation of an open-field site, at a distance of 3 m from a transmitting antenna, for frequencies ranging from 30 to 200 MHz.

APPENDIX

ANSWERS TO PROBLEMS

CHAPTER 2
2. 1 pF, 4 pF   **3.** 0.2   **4.** 2 μH, 2 μH, 1.5 μH   **5.** 0.75
6. (a) because no inductance is meaningful unless it is part of a closed current path, and the only possible current paths always include at least two inductances **(b)** because no inductance appears by itself in any formula for coupled noise

CHAPTER 3
6. $V_a = 2V_0 \cosh^{-1}[(s + k)/r_a] + V_k$, $V_b = -2V_0 \cosh^{-1}[(s - k)/r_b] + V_k$,
$V_0 = 1/\{2 \cosh^{-1}[(s + k)/r_a] + 2 \cosh^{-1}[(s - k)/r_b]\}$ volts,
$V_k = 1/2 - 1/\{1 + \cosh^{-1}[(s - k)/r_b]/\cosh^{-1}[(s + k)/r_a]\}$ volts, where k is as given
7. $C = 2\pi\epsilon l/\{\cosh^{-1}[(s + k)/r_a] + \cosh^{-1}[(s - k)/r_b]\}$
8. $V = \ln(1 + h^2/s^2)/[2 \ln(4h/d)]$ volts   **9.** $V = 4\pi\epsilon l/\ln(1 + h^2/s^2)$
10. $L = \mu l\{\cosh^{-1}[(s + k)/r_a] + \cosh^{-1}[(s - k)/r_b]\}/(2\pi)$

CHAPTER 4
1. parallel resonance occurs at $\omega^2 = 1/\{L_p[C_{1s} + C_sC_G/(C_s + C_G)]\}$
2. 0.133   **3.** transfer functions must be identical due to reciprocity
9. 40.5 dB   **10.** 0.95%   **11. (a)** 25.7 dB   **(b)** 44.7 dB
12. (a) 0.256 μH   **(b)** 1.64 μH   **13.** 1.51 μH   **14. (a)** 18.2 mA
(b) 1.54 mA   **(c)** 2.53 nWb   **15.** 15.6 dB

CHAPTER 5

1. the oscilloscope causes a stray return path that changes the current flow and the cross coupling   **2. (b)** 1.34%   **3. (b)** 1.77%   **4.** a signal current is not impeded from returning via other shields

CHAPTER 6

1. (a) audio **(b)** digital **(c)** RF **(d)** digital **(e)** audio **(f)** electromechanical **(g)** audio **(h)** RF **(i)** power conversion   **(j)** audio and (usually) digital
2. (a) audio to RF **(b)** digital to RF **(c)** digital to power conversion
4. 0.157 V   **5.** 266 V/m   **6.** because there would be a ground plane under a jumper, but not under a stitch   **7. (a)** 1 V **(b)** 0.393 V **(c)** 0.047 μF
8. 18.66 MHz

CHAPTER 7

1. $S(f) = \dfrac{1}{2\sqrt{2\pi}\,\sigma_f}\left\{\exp\left[-\dfrac{(f-f_0)^2}{2\sigma_f^2}\right] + \exp\left[-\dfrac{(f+f_0)^2}{2\sigma_f^2}\right]\right\}$

2 (a) $\frac{1}{2}\cos(2\pi f_0\tau)$ V^2 **(b)** yes, yes   **(c)** $\frac{1}{2}\cos(2\pi f_0\tau)$ V^2
(d) $\frac{1}{4}[\delta(f+f_0) + \delta(f-f_0)]$ V^2, where $f_0 = 10$ MHz   **6.** broadband

10. (a) $S(f) = \dfrac{V_0^2 T_c}{4}\left\{\delta(f) + \dfrac{\sin^2(\pi f T_c)}{(\pi f T_c)^2}\dfrac{\cos^2(\pi f t_r)}{[1-(2ft_r)^2]^2}\right\}$ **(b)** $1/(2t_r)$

(c) $S(f) \propto 1/f^6$, or 60 dB/decade

CHAPTER 8

2. 0.0183 μV/m/MHz$^{1/2}$, –34.7 dB$_{\mu V/m/MHz}$ **3.** 0.933 μV/m/MHz$^{1/2}$
4. –0.604 dB$_{\mu V/m/MHz}$ **5.** 11.6 μV/m, 21.33 dB$_{\mu V/m}$
6. 0.0373 μV/m/MHz$^{1/2}$, –28.56 dB$_{\mu V/m/MHz}$
7. 93.6 μV/m/MHz$^{1/2}$, 39.4 dB$_{\mu V/m/MHz}$

8. $|E_s(f_n)| = \dfrac{\sqrt{2}\,I_0 l \mu_0}{\pi r T_0}\left|\dfrac{\sin(\pi f_n t_1)\,\sin(\pi f_n t_r)}{\pi f_n t_r}\right| a_E$

CHAPTER 9

1. $\delta = 0.0209$ mm, 0.00209 mm, $t = 0.24$ mm, 0.024 mm
2. $\delta = 0.00295$ mm, $t = 0.034$ mm   **3. (b)** because the argument of the complex exponential is positive   **4.** 98 dB, 78 dB   **5.** 61 dB, 41 dB

6. 474.5 dB   **7.** 219.8 dB   **8. (a)** 126.8 dB   **(b)** 0 (since $A + R + M < 0$)
9. 189.4 ℧/m   **12.** 17.5 dB   **13.** 42.4 dB

CHAPTER 10
2. 60.02 dB   **3. (a)** 82.76 dB   **(b)** 65.76 dB   **4.** 58.83 dB   **5.** 48.19 dB
8. 68.25 dB   **9.** 80 dB   **10. (a)** 600 MHz   **(b)** attenuation would be zero
(c) 5 vertical, 2 horizontal

CHAPTER 11
1. 4.5 mV/m   **2.** 204 pF   **3.** 79.5 μH   **4. (a)** increase   **(b)** by a factor
of 15.6

CHAPTER 12
1. at 87 MHz (the third clock harmonic), $\|\mathbf{E}\| = 143$ μV/m, but the FCC limit is
100 μV/m

BIBLIOGRAPHY

BENNETT, WILLIAM R., and JAMES R. DAVEY, *Data Transmission*. New York, N.Y.: McGraw-Hill Publishing Company, 1965.

DAVIDSON, C. W., *Transmission Lines for Communications* (2nd ed.). New York, N.Y.: John Wiley & Sons, Inc., 1989.

ELLIOT, ROBERT S., *Antenna Theory and Design*. Englewood Cliffs, N.J.: Prentice Hall Press, 1981.

FINK, DONALD G., and DONALD CHRISTENSEN, *Electronic Engineer's Handbook* (3rd ed.). New York, N.Y.: McGraw-Hill Publishing Company, 1989.

"Guide for Construction of Open Area Test Sites for Performing Radiated Emission Measurements (ANSI C63.7-1988)," *American National Standards Institute (ANSI) Standards*. New York, N.Y.: The Institute of Electrical and Electronics Engineers, Inc., 1988.

HAYT, WILLIAM H., *Engineering Electromagnetics* (5th ed.). New York, N.Y.: McGraw-Hill Publishing Company, 1989.

JORDAN, EDWARD C., and KEITH G. BALMAIN, *Electromagnetic Waves and Radiating Systems* (2nd ed.). Englewood Cliffs, N.J.: Prentice Hall Press, 1968.

KRAUS, JOHN D., *Electromagnetics* (4th ed.). New York, N.Y.: McGraw-Hill Publishing Company, 1992.

"Methods of Measurement of Radio-Noise Emissions from Low-Voltage Electrical and Electronic Equipment (ANSI C63.4-1988)," *American National Standards Institute (ANSI) Standards*. New York, N.Y.: The Institute of Electrical and Electronics Engineers, Inc., 1989.

NARAYANA RAO, NANNAPANENI, *Elements of Engineering Electromagnetics* (3rd ed.). Englewood Cliffs, N.J.: Prentice Hall Press, 1991.

OTT, HENRY W., *Noise Reduction Techniques in Electronic Systems* (2nd ed.). New York, N.Y.,: John Wiley & Sons, Inc., 1988.

"Part 15—Radio Frequency Devices," in *Code of Federal Regulations, Volume 47: Telecommunications*. Washington, D.C.: U.S. Government Printing Office, 1991, subpt. A-B.

PLONUS, M. A., *Applied Electromagnetics* (2nd ed.). New York, N.Y.: McGraw-Hill Publishing Company, 1978.

RAMO, SIMON, JOHN R. WHINNERY, and THEODORE VAN DUZER, *Fields and Waves in Communication Electronics* (2nd ed.). New York, N.Y.: John Wiley & Sons, Inc., 1984.

SCHELKUNOFF, S. A., *Electromagnetic Waves*. Princeton, N.J.: D. Van Nostrand Company, Inc., 1943.

SCHWARZ, H. R., *Finite Element Methods*. San Diego, Calif.: Academic Press, Inc., 1984.

VANCE, EDWARD F., *Coupling to Shielded Cables*. New York, N.Y.: John Wiley & Sons, Inc., 1978.

INDEX

A (magnetic vector potential), 35–38, 149–50
Absolute magnitude, vector phasor, 133
Absorption loss, 168–70
Accuracy, EMI prediction, 4
Admittance, transfer, on antenna, 197–98
Air gap
  between shield layers, 187
  in common-mode choke, 225
Aircraft, EMI susceptibility, 2
Aluminum, shielding properties, 170–71
Ambient noise at test sites, 239–40
Ampere's law, 36, 48, 189, 191
Amperes^2, 107
Amplifier
  audio, 82
  differential, 65, 69–70
  emitter-follower, 94
Amplitude-phase form of Fourier series, 105
Anechoic chamber (electromagnetic), 239–40
Antenna
  biconical, 234–35
  dipole, 194–200
  dipole, tuned, 232, 234
  monopole, 216–18
  slot, 194–97, 201
Antenna factor
  biconical, 234–35
  tuned dipole, 234
Aperture current, 191–94

Approximations, asymptotic
  capacitive coupling, 19
  inductive coupling, 32
Arc spray process, 207
Area between signal and ground leads, 55–56
Area of current loop, 142, 150–52, 155
Areas, component layout, 84, 92
Array of slots, 204
Attenuation
  bypass capacitor, 218, 220
  cable shield, 53–55, 62, 65
  noise coupling path, 6, 7, 12–13
  power bus filter, 102
  RF choke, 222
  shield enclosure, 167–70, 176–81, 186
  shield enclosure, slotted, 201–7
  site (test), 237–39
Attenuation constant (α), 168
Audio, circuit class, 82–83
Autocorrelation function
  digital random process, 117–19
  periodic function (time), 107–8
  random process (ensemble), 109–10
Automotive ignition system, 2, 5
Average of multiple noise sources
  coherent current, 147
  coherent voltage, 137–38
  incoherent current, 148
  incoherent voltage, 140–41